1979

P9-ASI-371

The History, Biology, Damage, and Control of the Gypsy Moth

Silhouette of the male gypsy moth, taken in Centre County, Pennsylvania. *Courtesy: D. A. Noviello.*

The History, Biology, Damage, and Control of the Gypsy Moth

Porthetria Dispar (L.)

Michael H. Gerardi and
James K. Grimm

Rutherford • Madison • Teaneck
Fairleigh Dickinson University Press
London: Associated University Presses

Associated University Presses, Inc.
Cranbury, New Jersey 08512

Associated University Presses
Magdalen House
136–148 Tooley Street
London SE1 2TT, England

Library of Congress Cataloging in Publication Data
Gerardi, Michael H.
    The history, biology, damage, and control of the gypsy moth,
Porthetria dispar (L.)
    Bibliography: p.
    Includes index.
    1. Gypsy-moth.   2. Gypsy-moth—Control.
I. Grimm, James K., joint author.   II. Title.
SB945.G9G47   634.9′6′781   76–20321
ISBN 0–8386–2023–X

The authors thank the National Academy of Sciences for permission to
reprint Model 2 and Model 4 from *Insect-Pest Management and Control,
Principles of Plant and Animal Pest Control,* Volume 3, Publication ISBN
0–309–01695–9, Committee on Plant and Animal Pests, National Academy
of Sciences-National Research Council, Washington, D.C., 1969.

The authors thank *Science* for permission to reproduce Fig. 1 and Tables 1
and 3, from "Gypsy Moth Control with the Sex Attractant Phermone,"
Beroza, M. and Knipling, E. F., *Science* 177 (1972): 19–27. Copyright by
the American Association for the Advancement of Science.

Contents

Illustrations

Preface

Not since 1896, with the publication of *The Gypsy Moth* by
E. H. Forbush and C. H. Fernald, has a text been written
on the biology of the gypsy moth; and since the insect's
introduction into the United States over 100 years ago, the
economic and social impact of the gypsy moth in the northeastern United States has increased tremendously from year
to year. For these two reasons we have prepared from
numerous individual, government, and journal publications
and correspondence a comprehensive and simplified survey
of the history, biology, and control of the gypsy moth.

Since 1896 advances and failures in the control of this
insect pest have occurred, and a general study of the
research and literature pertinent to the gypsy moth appears
to be needed. This text attempts to update and present the
basic and essential information concerning the history,
biology, and control of the gypsy moth.

In addition, the text also provides a general study in pest
management and control for beginning entomology students.

Acknowledgements

The authors would like to thank the following individuals for their reprints and correspondences, which were most helpful in preparing this manuscript: N. C. Anderson, A. A. App, J. L. Bean, M. Beroza, B. A. Bierl, J. H. Bitzer, E. A. Cameron, R. W. Campbell, R. T. Cardé, H. C. DeRoo, C. I. Doane, D. M. Dunbar, A. J. Forgash, R. R. Granados, J. Granett, R. A. Hamlen, J. B. Hanson, J. T. Katsanos, J. D. Kegg, E. F. Knipling, D. E. Leonard, F. B. Lewis, W. A. Martin, H. Marx, J. D. Nicholas, J. L. McDonough, W. A. Merriam, E. C. Paszek, F. M. Phillips, J. D. Podgwaite, J. W. Quimby, R. C. Reardon, L. D. Rhoads, J. V. Richerson, F. E. Sandquist, C. P. Schwalbe, E. E. Simons, Z. Smilowitz, G. R. Stephens, L. J. Stevens, C. R. Sullivan, J. G. R. Tardif, A. D. Tomlin, G. C. Tower, N. C. Turner, D. R. Wallace, G. S. Walton, R. M. Weseloh, and W. C. Yendol.

We are indebted also to many industries and government departments and their agencies. Acknowledgment is made of the assistance received from Abbott Laboratories, Baychem Corporation, International Minerals and Chemical Corporation, National Academy of Sciences, New Jersey Department of Agriculture, Northeastern Forest Experiment Station, Pennsylvania Department of Environmental Resources, Sandoz Corporation, Union Carbide Corporation, Unted States Department of Agriculture, United States Forest Service, and World Health Organization.

Photographs are used with permission from R. A. Hamlen, D. A. Noviello, Unted States Department of Agriculture, and United States Forest Service. Illustrations and other original materials have been provided by B. A. Bierl, M. Beroza, Canadian Entomologist, Entomological Society of America, J. B. Hansom, International Minerals and Chemical Corporation, D. E. Leonard, Massachusetts Forest and Park Association, J. L. McDonough, W. A. Merriam, National Academy of Science, Pennsylvania Department of Environmental Resources, R. C. Reardon, Sandoz Corporation, Science, C. R. Sullivan, C. C. Towers, United States Department of Agriculture, and D. R. Wallace.

Deep appreciation is given to M. A. Black and D. W. Gaines for their assistance in perparing this manuscript.

Introduction

In recent years in the northeastern United States, several insects destructive to forest and shade trees have become a major menace, as well as a threat to neighboring woodland areas. The damage caused by one insect in particular, the gypsy (gipsy) moth, *Porthetria* disper Linnaeus (Lepidoptera: Lymantriidae), has been increasing yearly on a geometric scale throughout many parts of its range in the Northeast. Because of this progressive increase in damage and current inability to achieve effective, large-scale control against this insect, many consider the gypsy moth to be the most serious pest of forest, shade, and fruit trees in the northeastern United States. The gypsy moth's rise to notoriety was caused by several factors: the absence of effective parasites and predators, the abundance of favored food, and the presence of numerous ecological factors favoring the survival of large numbers of gypsy moth caterpillars have allowed the insect to reach impact population levels, resulting in millions of acres of defoliation and millions of dollars in damage.

In 1953 the gypsy moth defoliated over 1,500,000 acres of woodlands in New England, New Jersey, and throughout the eastern parts of New York and Pennsylvania. In 1967, 52,000 acres of timber land in New England were defoliated; in 1968, 80,000 acres were defoliated; in 1959, 225,000 acres of defoliation resulted; and in 1970, over 900,000 acres of woodlands in the northeastern states were defoliated by gypsy moth caterpillars (Koski, 1971a). A high of 2,000,000 acres

13

of defoliation throughout the Northeast occurred in 1971. Some of this acreage sustained second and even third defoliations by the gypsy moth. As the moth spreads from the Northeast, individual states are suffering from record-high gypsy moth defoliations. In Pennsylvania nearly 900,000 acres of forest lands were attacked by the insect in 1973 (Marx, 1974b).

If the gypsy moth is not controlled, over 110,000,000 acres of woodlands in the northeastern states may sustain substantial losses. An additional 100,000,000 acres of valuable Ozark and Appalachian hardwood forests also could be damaged if the insect is allowed to extend its range through Pennsylvania toward the South and West. The gypsy moth presently infests 12 states and continues to increase its range each year. States already infested include Maine, Vermont, New Hampshire, Connecticut, Rhode Island, Massachusetts, New York, New Jersey, Pennsylvania, Delaware, Maryland, and Michigan. In addition to these states, the southern portions of two Canadian provinces, Quebec and Ontario, are also infested with the gypsy moth. Although attempts are being made to prevent the spread of this insect, the gypsy moth is expected to extend its range from the generally infested Northeast south to Florida and west to the Mississippi River.

Aside from the insect's environmental and economic impact, more legislation and money have been used in attempts to control the gypsy moth than any other insect pest in the United States. The largest biological control program ever attempted by the United States Department of Agriculture against an insect pest is presently being conducted against the gypsy moth. Over 40 plant species have been identified as primary food sources for this insect; and over 500 additional plant species have been observed to serve as secondary food sources (Metcalf et al., 1962). This large number of host plants allows the gypsy moth to establish large populations, to overcome many applied pest-control measures, and to expand its range throughout the eastern United States and southeastern Canada. Attempts to control this insect have proved only partially successful. Its present and potential threat to the Northeast, Appalachian, and Ozark hardwood forests indicates that much attention is due the history, biology, and control of the gypsy moth.

A review of the available literature reveals that the gypsy moth is classified in the families Lymentriidae, Liparidae, and Orgyidae, and in the genera *Porthetria, Lymantria, Liparis,* and *Ocneria.* Its habitat is located in the temperate regions of the world, extending throughout central and couthern Europe, northern Africa, and central and southern Asia, including Japan and Ceylon. This range extends from a latitude of 20° north in southeastern Asia to 58° north in Scandinavia (Leonard, 1974). The gypsy moth is also present in Great Britain, although considered a rare species. In North America the gypsy moth is an introduced species, brought to the United States in 1869.

The insect is cyclic in nature with peak populations occurring at irregular intervals. The gypsy moth is considered a major pest to people, forests, and orchards in Yugoslavia, France, Rumania, and Japan, where population peaks occur at 2–5, 4–5, 9, or 9–10 year intervals (Marx, 1973b). Although it is held in check to some degree in Europe and Asia by parasites and predators, it has proved to be as destructive a pest there as in North America, where many of these natural enemies are absent. In the northeastern United States the gypsy moth is the most dangerous defoliator of hardwoods and evergreens and the most serious threat to the oak forest.

The gypsy moth continues to dominate forest pest-control activities. Pennsylvania is now the crucial buffer state for gypsy moth control work, and the United States

Forest Service estimates that two-thirds of the state's 15,000,000 acres of commercial forests are highly susceptible to defoliation by this insect. Since Pennsylvania is strategically located between the generally infested Northeast and the open woodlands to the south and west, great emphasis has been placed on halting the spread of the gypsy moth in this state. However, the insect has not been stopped in Pennsylvania; rather, it is slowly extending its range into the southern hardwood forests.

Population increases in the gypsy moth are slow in the New England states where oak, its favored host, is a minor component of the woodlands. With forest compositions more favorable to gypsy moth population growth in Pennsylvania, New York, New Jersey, and states farther south, and with more favorable climatic conditions in these woodlands, the full impact of the insect is beginning to reveal itself, accompanied by rapid population increases.

Although some conservationists believe that the threat of the gypsy moth is overrated and the insect is incapable of killing one in twenty trees, this may be true only in the older, more resistant forest types in northern New England (Marx, 1970a). However, in the major infestations outside northern New England, the impact of the gypsy moth cannot be underrated. According to James Nichols, Chief of Division of Forest Pest Management, Pennsylvania Department of Environmental Resources, the gypsy moth cannot be controlled with present technology, and the resulting impact of the insect is expected to be many times greater than that of the chestnut blight (Nichols, 1971).

Table 1 - 1

ACRES DEFOLIATED

Statistics on Major Forest Insect Defoliators of Pennsylvania
1965–1973

Year	Gypsy Moth *Porthetria dispar*	Saddled Prominent *Heterocampa guttivitta*	Fall Cankerworm *Alsophila pometaria*	Forest Tent Caterpillar *Malacosoma disstria*	Oak Leaf Roller *Archips semiferanus*	Oak Leaf Tier *Argyrotoza semipurpurana*
1965	0	0	522,000	8,700	0	608,000
1966	0	1,100	825,200	0	0	611,500
1967	0	66,000	715,500	700	240,000	786,500
1968	60	83,000	0	0	306,200	448,500
1969	830	90,600	0	6,500	247,400	0
1970	10,500	70,000	0	100,000	985,500	0
1971	92,200	46,000	0	204,800	1,045,100	0
1972	404,060	15,500	0	25,800	609,400	59,200
1973	856,710	0	9,700	50	113,000	9,050
1974	479,590	0	7,250	0	235,600	0
1975	317,880	0	363,000	0	1,750	0

(Courtesy: Simons, 1973; and Pennsylvania Department of Environmental Resources, 1975)

Table 1 - 2

GYPSY MOTH DEFOLIATION 1972–1974

ACRES

STATE	TOTAL 1974[a]	TOTAL 1973[a]	TOTAL 1972[b]
Connecticut	120,980	333,215	513,880
Maine	860	490	40
Massachusetts	76,903	43,970	20,480
New Hampshire	*	30	200
New Jersey	28,102	254,865	226,140
New York	42,350	248,441	177,605
Pennsylvania	479,590	856,710	404,060
Rhode Island	2,120	35,925	22,510
Vermont	*	200	4,215
TOTAL	750,905	1,773,846	1,369,130

* Not available
[a] Source: Marx 1974b
[b] Source: Marx 1972c

The History, Biology, Damage, and Control of the Gypsy Moth

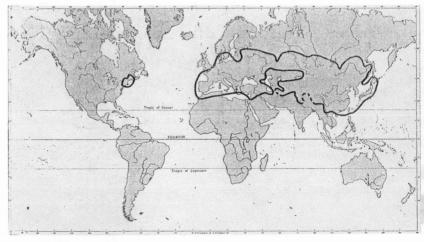

Distribution Map of the Gypsy Moth
Present range of the gypsy moth is enclosed in the heavy black lines. Countries in which the insect is known to occur are: Austria, Belgium, Bulgaria, Czechoslovakia, Denmark, France, Germany, Great Britain (rare), Greece, Hungary, Italy, Netherlands, Poland, Rumania, Spain, Sweden, Switzerland, and Yugoslavia in Europe; Canada and United States; Afghanistan, China, Formosa, India, Iraq, Israel, Japan, Kashmir, Korea, Lebanon, Syria, Turkey, and U.S.S.R. in Asia; and Algeria, Morocco, and Tunisia in Africa. *Anon 1953*

1

Spread of the Gypsy Moth

Because Confederate cotton was unavailable during and shortly after the Civil War, the Union States were forced to seek a new source of fiber; in the spring of 1869 M. Leopald Trouvelet, an American naturalist, brought a number of gypsy moth egg clusters from southern France to Medford, Massachusetts. Trouvelet had hoped to produce a commercial source of silk by developing a hardy race of silk-producing insects, crossing the gypsy moth with the silkworm moth, in order to control "wilt" disease (or flacheria), then causing severe problems in some silkworm industries.[1] However, during the course of his experiments some of the eggs were lost, and some of the caterpillars escaped from his home at 27 Myrtle Street. Although this accident was made public at the time, it did not receive much attention even though the gypsy moth was immediately recognized as a pest. This accident and a combination of circumstances—lack of abundant natural enemies and favorable climatic conditions—allowed the insect to gain a foothold in North America.

By 1880 the gypsy moth had infested approximately 400 square miles around Medford; but it was not until the summer of 1889, twenty years after its introduction to the United States, that the insect became so abundant and destructive that it arrested the general public's attention. At that time the gypsy moth was completely defoliating fruit and shade trees over approximately 360 square miles; and the caterpillars, unlike native woodland pests, invaded several towns, attacking shade trees and becoming a public nuisance. The reaction to the insect was so serious that the state of Massachusetts appropriated funds for its extermination by the State Department of Agriculture. This act resulted in the first state law in the United States requiring the extermination of an insect pest and thereby established a legal precedent to enter private property for the purpose of eradication or control of a pest insect. The areas treated for gypsy moth infestations were centered around Boston. Treatment was conducted from 1892 to 1898 and consisted of lead arsenate developed and manufactured by the Federal Bureau of Entomology for use against the gypsy moth in 1892. By the summer of 1898 little defoliation could be seen in the infested areas, and field workers believed that the gypsy moth had been successfully exterminated in these areas. In 1900 the state legislature of Massachusetts, believing that the gypsy moth eradication program had succeeded in bringing the pest under control, ordered the program discontinued. Much of the early work against the gypsy moth was conducted from the Gypsy Moth Laboratory at Melrose Highlands, Massachusetts.

Between 1900 and 1905 populations of the insect increased enormously from year to year, with a corresponding increase in the amount of damage caused by its larvae. The gypsy moth gradually expanded its territory into neighboring New England states. Thousands of acres of woodlands were defoliated, and many trees in several residential sections were injured. The situation became critical; and in 1905 the Massachusetts Department of Agriculture resumed control work. By this time over 2,200 square miles of woodlands were infested. For this first time infested forests could be found outside of Massachusetts—in Maine, New Hampshire, and Rhode Island.

With the spread of the insect into neighboring states, federal funds were appropriated by Congress in 1906, and the Secretary of Agriculture was authorized to take necessary measures, along with the states involved, to prevent further spread of the gypsy moth. Due to the rapid spread and enormous population size of the insect, state and federal agencies could only treat some of the infested residential areas and only slightly inhibit its further spread.

The gypsy moth was imported to the United States a second time; however, this importation was accidental. A New Jersey State Agricultural Inspector found egg clusters of the gypsy moth on a large estate near Somerville in 1920. This infestation occurred in a large plantation of blue spruce trees, *Picea pungens,* several acres of which were completely defoliated. Records revealed that these egg clusters had been imported from the Netherlands with the trees in 1910. This importation had occurred two years prior to the enactment of the Plant Quarantine Act, thus demonstrating the importance of such legislation.

On August 20, 1912, Congress passed the Plant Quarantine Law, which prevented the shipment of any insect life stage from infested areas to noninfested areas. The first federal gypsy moth quarantine was established from this law and became effective on August 1, 1913. Much of the Northeast was placed under quarantine; since then, shipments of regulated materials from infested to noninfested areas have been under federal quarantine and regulation.

Despite state and federal control measures, the gypsy moth continued to spread from isolated areas of infestation in New England and from the generally infested areas in Massachusetts. By 1914 the gypsy moth was scattered throughout the entire southern half of New Hampshire and had expanded its range east to Bangor, Maine and west across the Connecticut River in Massachusetts and into Vermont. Towns in eastern Connecticut were found to be infested with the insect. During the First World War, with reduced control measures in effect due to lack of available manpower, equipment, and money, the gypsy moth continued to spread throughout the New England area. By 1922 isolated infestations were found farther west in Vermont, Connecticut, Massachusetts, and New York, where the gypsy moth first appeared along the Massachusetts State Line. By 1922 more legislation had been passed in the New England States concerning the control of the gypsy moth than of any other pest.

In order to control the spread of the gypsy moth further, a conference was held in Albany, New York, on November 26, 1922, bringing together representatives from all infested states, the United States Department of Agriculture, and Canada. From this conference, state and federal funds were appropriated in 1923 for the creation of a gypsy moth control zone to prevent the westward spread of the insect. This barrier zone, located east of the Hudson River, extended 250 miles from Long Island Sound, New York, to the Canadian border. Varying in width from 25 to 30 miles, depending on the terrain through which it passed, the zone enclosed approximately 8,000 square miles. This area was selected as the smallest and most feasible area in the Northeast that could be maintained to prevent nation-wide spread of the insect. All infested areas east of the control zone were to be treated by the states involved, aided by liberated, imported parasites and predators of the gypsy moth provided by the Federal Bureau of Entomology. For the most part treatment consisted of scouting and clean-up work, with the application of lead arsenate in infested woodlands. Control work in New York was financed by the State, with assistance from the USDA.

In 1924 more infestations were found farther west in

Massachusetts and Vermont, and for the first time a colony of gypsy moths was found in Canada at Henrysburg, Quebec. With these new infestations, the barrier zone was moved west to include the entire state of Vermont and additional towns in northwestern Connecticut, and a Canadian quarantine was enacted to cover the southern towns in Quebec. At this time the federal quarantine area consisted of the entire states of Massachusetts, Rhode Island, and Connecticut, and the major portions of Vermont, New Hampshire, and Maine. Prior to the establishment of the barrier zone, isolated infestations had been found in New York (1912), Ohio (1914), New Jersey (1914), and Pennsylvania (1920). These isolated areas indicated the potential threat to distant woodlands by the transportation of the gypsy moth and the future ineffectiveness of the barrier zone against such transportation. Most of these infestations were eradicated quickly; however, some took as long as 15 years, allowing further spread of the insect. By 1927 over $25,000,000 had been spent on eradication and control programs against the gypsy moth.

During 1929 and 1930 the number of infestations further increased in the barrier zone and in the states already infested. Congress appropriated additional funds, and much of the territory previously classified as lightly infested was reclassified as generally infested. With the Melrose Laboratory defunct in the early 1930s, gypsy moth control work and experimentations were centered in the Northeastern Forest Experimentation Service, USFS, in New Haven, Connecticut.

In 1933 after the passage of the National Industrial Recovery Act, once again funds were appropriated for scouting and clean-up of the woodlands and for lead arsenate treatment between the barrier zone and the Connecticut River. Those agencies involved in this major undertaking consisted of the State Forest Services of the infested states, the Civilian Conservation Corps, the Federal Bureau of Entomology and Plant Quarantine, the U.S. Department of Agriculture, and the U.S. Department of the Interior.

In 1934 defoliation continued in enormous proportions throughout New England, especially in the northern third of Massachusetts near Princeton and New Salem. More forest land and additional towns throughout Maine, New Hampshire, Vermont, and Connecticut were added to the quarantine area during this year.

Major and minor peaks of defoliation in New England appeared in 1915–1917, 1921–1922, 1926–1928, and 1934–1935. Damage caused by the gypsy moth in New England amounted to 3,720,500 acres defoliated from 25 to 100 percent over the 12-year period from 1925 to 1936 (Burgess and Baker, 1938). The rate of defoliation during this time period was approximately 310,000 acres per year. From 1894 to 1936 over 5,500,000 acres of northeastern woodlands had been defoliated; and from 1906 to 1934 the federal government alone spent over $40,000,000, an average expenditure of over $1,000,000 per year for treatment and control of the gypsy moth.

In 1932 an infestation located in an area of 400 square miles was discovered in the Luzerne and Lackawanna Counties of Pennsylvania. This large, isolated infestation west of the barrier zone indicated that the gypsy moth had gained a foothold beyond the control area. Through the use of lead arsenate this area was kept under control for ten years, but it was gradually absorbed into the expanding range of the gypsy moth.

During World War II insufficient manpower, transportation of the gypsy moth life stages, high winds, and reservoir populations in the inaccessible and untreated woodlands in the Adirondacks of northern New York, the Delaware River Valley, the Green Mountains of Vermont, and the Catskills of western Connecticut enabled the gypsy moth to spread west and south from the generally infested New England states into New Jersey, eastern New York, and Pennsylvania.

Many of these locations, such as the Delaware River Valley in New York and central and southern portions of New Jersey, remain major trouble spots for gypsy moth control work today.

In 1945 the gypsy moth defoliated over 800,000 acres of woodlands. From 1945 to 1963 the extent of its damage and spread was controlled to varying degrees with the use of DDT. In 1963 with the ban of DDT and use of less persistent and less effective pesticides, the gypsy moth began to increase its rate of westward and southward expansion across New York and Pennsylvania.

From 1969 to 1970 the gypsy moth was on the increase in the Northeast. The heaviest growth of gypsy moth populations *ever experienced* in Connecticut occurred in 1970 (Marx, 1970a). The overall gypsy moth picture for the early 1970s indicated an upswing in population throughout the Northeast and along the periphery of the generally infested area. For the first time the insect appeared west of the Susquehanna River in 1969, and in 1971 it was known to be present in 48 of the 67 counties of Pennsylvania. By 1973 the insect had spread to 61 of the state's counties, and today the entire state of Pennsylvania is considered to be infested with the gypsy moth. The insect has caused heavy defoliation while moving across the state. Much of this defoliation has followed a geometric progression: 800 acres defoliated in 1969; 10,000 in 1970; 100,000 in 1971; and 900,000 in 1973. In 1976 a total of 756,000 acres were defoliated with the resurgence of the gypsy moth after infestations had largely collapsed following severe defoliation in 1973 (Nichols, 1976). From 1970 to 1975 the gypsy most has destroyed almost $8.5 million in timber and pulpwood alone in Pennsylvania (Pa. Dep. Envir. Res., 1975). The insect continued to push through Pennsylvania into

West Virginia and Maryland. By 1971 the male moth had been found in all Maryland counties except: Allegany, Garrett, and St. Mary's.

The insect is also increasing its range throughout Canada; however, the rate of spread through Canada is much slower than in the United States. By 1958 a continuous population of the insect was established along the St. Law- rence River, from the New York and Vermont State borders west to within a few miles of Ontario (Brown, 1968). The gypsy moth was found in southeastern Ontario for the first time in 1969 (Rose, 1969). Egg masses were found on Wolfe Island near Kingston, and by 1970 infestations were located on Howe Island and on the mainland of Ontario. Over 13,000 acres near Kingston were infested in 1971 (Marx, 1972a). A large variety of host plants in southern Ontario and Quebec has caused the insect to be considered a serious pest in Canadian forests. This threat to Canadian forests appears more menacing as the insect continues to spread across Canada. In 1971 male moths were captured in New Brunswick and Nova Scotia (Brown, 1973). However, if a cold-hardy race of the gypsy moth does not evolve, it may reach its environmental limits in southern Quebec and eastern Ontario (Marx, 1972c).

Since 1869 the natural spread of the gypsy moth larvae has progressed at the rate of eight to ten miles per year. By 1974, 105 years after the introduction of the gypsy moth into the United States, the insect had increased its range over 200,000 square miles of the United States and Canada, and over $110,000,000 had been spent by the federal government and millions more by state governments in attempts to combat this pest. Other states are being invaded, and it is expected that the gypsy moth eventually will spread throughout the oak regions east of the Great Plains.

The potential spread of the gypsy moth has received

renewed interest because of recent trapping and quarantine programs. In 1973 the male moth had been trapped in Ohio, West Virginia, Virginia, North Carolina, Georgia, Florida, Tennessee, Indiana, Illinois, and California; and the gypsy moth either in egg, larval, or pupal form had been transported accidentally outside of the generally infested Northeast to Alabama, Minnesota, Missouri, Texas, and Wisconsin.

Because of present and potential gypsy moth spread and damage, efforts are constantly being made to develop better means of control and survey. In addition to these efforts, the potential threat of the insect to southern and midwestern forests is being analyzed; and on the basis of stand compositions, the gypsy moth is expected to occupy the entire region of the eastern United States, in which oaks are a major component of the forest stands (Perry, 1955).

The impact of the gypsy moth in southern forests is expected to have a greater effect than the present one in the northeastern forests. This is due to the presence of more suitable hosts in the southern forests and the warmer, southerly climate that favors earlier egg hatching and slower larval development. With earlier hatching and slower development, the southern hardwood forests would be exposed to longer periods of larval feeding. In addition, the possibility exists for a second generation of the insect each year in the more southerly woodlands. The impact of the insect will affect not only timber resources, but also recreational values and tourist industries. Southern areas, with a high percentage of susceptible hosts now being invaded, are the extensive oak forests of the Appalachian and Ozark Mountain ranges and the southern oak-pine forests. The greatest impact of the gypsy moth will be felt in oak forests and primarily on the ridge and plateau areas that support large stands of chestnut oak and white oak.

NOTE

1. "Wilt" disease is a viral infection causing a swelling of the larval body and is often called jaundice for the yellowish color of infected larvae.

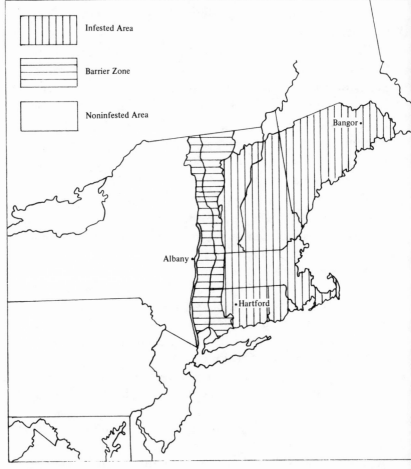

Gypsy Moth Barrier Zone, 1923–1934. *Courtesy: USDA, after Burgess and Baker, 1938*

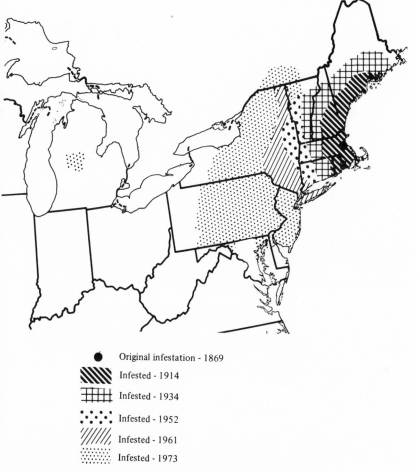

●   Original infestation - 1869

  Infested - 1914

  Infested - 1934

  Infested - 1952

  Infested - 1961

  Infested - 1973

Gypsy Moth Spread 1869—1973. *Courtesy: USDA, 1973a*

2

Life Stages

The gypsy moth is an insect that illustrates complete or complex metamorphosis and is listed with those insects referred to as Holometabola. The life history of the insect is divided into four distinct stages: (1) the egg stage; (2) the larva or feeding stage; (3) the pupa, a quiescent transformation stage; and (4) the adult or reproductive stage. These stages reveal changes in form, structure, and food preference during the insect's development to maturity.

The first stage in the development of the gypsy moth is the egg. Hatching from the egg, the larva or caterpillar appears, showing little resemblance to the adult. During the growth of the larva, its "skin" or cuticle is shed four or more times. This process of shedding is known as molting or ecdysis, and the cast skin is called the exuvium. Time intervals between molts are known as stadia, and the form assumed by the insect during a particular stadium is called an instar. The first instar appears when the insect hatches from the egg, and at the end of this stadium the first molt occurs; then the insect assumes its second instar, and so on. Ecdysis is regulated by the periodic release of certain hormones. Before the final molt, the larve spins a cocoon and pupates, transforming itself from an active larva to a resting pupa. From the pupa, the adult moth or imago will emerge, thus completing the process of metamorphosis.

EGG STAGE

The egg of the gypsy moth is globular, white or transparent, and slightly more than 1 mm (1.10–1.25 mm) in diameter. Eggs are deposited in several layers and are laid in clusters or masses of 100 to 800 eggs each, commonly 400 to 500 each. The masses vary greatly in size, compactness, surface area, and shape. Egg masses range from 1.3 cm to 5.1 cm, but the majority are roughly a 2.5 cm by 3.8 cm oval, raised in the center. The egg mass is covered with velvety, buff-colored hairs from the body of the female moth. This covering not only protects the eggs from excessive evaporation, but also makes the egg mass unattractive to some birds. The egg clusters look and feel like pieces of chamois cloth.

The size of the egg clusters is affected by population density and food supply experienced by the parental generation. In light infestations where food is plentiful, clusters are larger than the average, and in heavy infestations where food is limited or scarce during larval development, the resulting egg clusters are smaller than average. A direct relationship exists between pupal weights of the females and the number of eggs produced by the females emerging from these pupae, that is, heavier pupae yield larger females, which lay more eggs (Maksimovic, 1958).

The female normally lays all her eggs in one mass. This mass is oblong, tapering at one end. In a cross-sectional view, the egg mass appears to be layered, containing five or more planes of eggs. With the female resting on a hard surface, the broader portion of the mass is laid first. Then she slowly moves forward from this position until egg laying is completed with the formation of the tapered end of the mass. The eggs in the broad portion of the clusters are formed first in the polyrophic ovaries, while those eggs formed last in the ovaries are deposited in the tapered region of the cluster.

Based on the general shape of the egg mass, the mass can be divided into three regions: the basal portion, the central portion, and the tapered portion. The basal portion contains those eggs that are usually larger than the eggs in the remaining two portions. The larvae that hatch from these eggs undergo additional molts less readily than the larvae from the remaining eggs (Leonard, 1970a). Since the egg mass is raised in the center, the central portion contains more eggs than either of the other two portions. The tapered end contains the smallest size eggs. This reduction in size is due largely to a yolk deficiency in the eggs. In addition, the fewest number of eggs is found in the tapered portion of the egg mass; most of the nonviable eggs laid by the female are also found in this section.

When parasitism of the egg mass occurs, the majority of the eggs that are parasitized are contained in both the basal and the tapered portions of the egg mass. Since more eggs are in the central portion, this section statistically would have a lower percentage of parasitism than the basal or tapered portions. Consequently, as the size of the egg mass decreases, the efficiency of parasitism increases (Dowden, 1961).

LARVAL STAGE

It is during the larval stage of development that the gypsy moth inflicts all of its damage.

The mature larva is hairy, 3.8 cm to 6.4 cm long, illustrating eruciform body structure. The vesicular and acuminate hairs originate from spots or tubercles on all segments of the caterpillar. Coloration of late and early instars is similar. The head has yellow markings, and the body is dusky or slate colored, peppered with numerous small and darker spots. There are three light stripes along the back. From the head, there are five pairs of blue spots, followed by six pairs

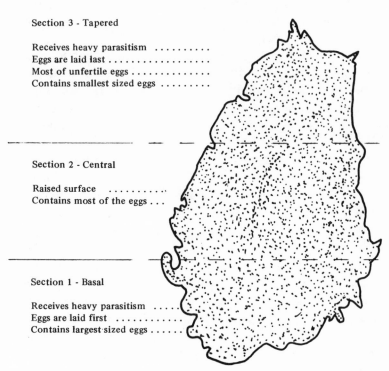

Section 3 - Tapered

Receives heavy parasitism ....
Eggs are laid last ....
Most of unfertile eggs ....
Contains smallest sized eggs ....

Section 2 - Central

Raised surface ....
Contains most of the eggs ...

Section 1 - Basal

Receives heavy parasitism ....
Eggs are laid first ....
Contains largest sized eggs ......

Characteristics of the Gypsy Moth Egg Mass

Gypsy Gypsy Moth egg mass. *Courtesy: US Forest Service*

of brick red spots. These spots, one pair per segment, are located on segments 2 through 12 and show more prominently as the larva approaches maturity. This double row of colored spots is the most distinguishing characteristic of the larva. The legs, prolegs, and general ventral surface of the caterpillar are light grayish-brown in color. The larva has mandibular or chewing mouthparts.

A newly hatched larva is about .2 cm long and passes through a series of molts or instars before becoming fully developed. In North America, the male larva normally has five instars, and the female larva normally has six instars. An increase in the number of instars for each sex is possible when the available food supply is reduced or when larvae become crowded. More quickly developing larvae tend to be heavier, with a higher reproductive potential than more slowly developing larvae.

According to Goldschmidt (1934), there are five types or geographic races of *Porthetria dispar* throughout the world. These races are as follows:

(1) both sexes with five instars each
(2) both sexes with six instars each
(3) all males with five instars and all females with six instars
(4) all males with five instars and females with five or six instars
(5) all females with six instars and males with five or six instars

Goldschmidt (1934) considered the variations and number of molts undergone by a gypsy moth larva to be dependent upon a set of multiple alleles, M_a, M_b, and M_c. Allele M_a causes four molts in both sexes; allele M_b causes four molts in the male and five molts in the female; and allele M_c causes five molts in both sexes. The race of gypsy moth found in North America and in parts of Europe is characterized by all males having five instars and all females having six instars. Sex-determining factors for the gypsy moth are as in other Lepidopterans. The female is the heterogametic sex, with her determining factors conceived of as cytoplasmic in origin, while the male-determining factors lie in the X-chromosome.

A simple, whole-mount, microscopic technique has been devised by Levesque (1963) for determining the sex of fully developed embroys and early-instar larvae of the gypsy moth. In this technique, alcohol-preserved larvae are treated through a series of chemical staining solutions, and the gonads are removed. Under high-power magnification, the contrasting characteristics of the male and female gonads show up clearly. In the male, each testis consists of four well-defined, kidney-shaped lobes from the fully developed embryo stage through the late instars. In the female, the ovaries are undifferentiated in the embryo and early first instar larvae. Toward the end of the first instar, three definite, oval-shaped lobes are formed, which are connected by a stalk-like outgrowth, the calyx. This calyx is absent in the male gonads.

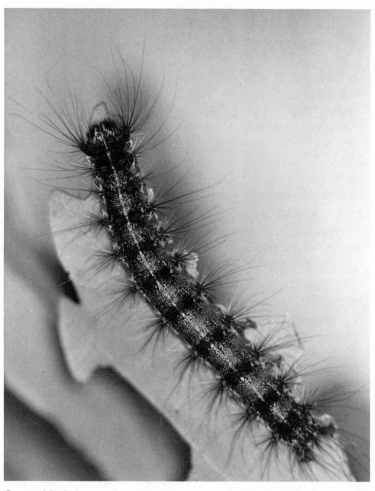

Gypsy Moth larva. *Courtesy: Pennsylvania Dept. of Environmental Resources*

PUPAL STAGE

When fully grown, the larva sheds its cuticle and evolves into the pupa. The pupa or cocoon is reddish-brown in color and loosely enclosed by a few strands of silk. The naked pupa is fastened with silk to an object. There are a few short, light-brown or reddish hairs around the spiracles and across the thoracic and abdominal segments. Female pupae are larger than male pupae; the average length of the female is 2.5 cm, and the average length of the male is 1.9 cm.

Gypsy Moth pupa. *Courtesy: USDA*

ADULT STAGE

The adult moths are different in appearance, the sexes displaying dimorphism. Adult sex ratios in a given population may vary from 80% female moths to only 2% female moths (Campbell, 1963). Distortion of the sex ratio in a population is caused by disease and desiccation. Such is the case during instars four, five, and six and among prepupae, where these two factors strongly select against the female (Campbell, 1963). Parasitism also selects against the female.

MALE MOTH

The male has a slender, dark-brown body and light to dark-brown wings with black markings. Blackish, wavy bands are present across the forewings, with arrowhead markings near the anal edge of the wings. These markings are absent on the hindwings. 2.5 cm to 3.8 cm. The antennae of the male are pectinate in appearance and aid in locating the flightless female. They are broad and light brown with a tufted mass of hair at the base, directly behind the head. The mouthparts are in the form of a coiled sucking tube and are not used for feeding.

FEMALE MOTH

The female moth, having a wingspread of 5.0 cm to 6.4 cm, is much larger than the male. She is white or grayish white with a heavy, cylindrical abdomen clothed in yellow hairs. Present on the forewings are brown or blackish bands and arrowhead markings, and a marginal transverse line of dark-colored dots appears near the outer border of both fore- and hindwings. The antennae of the female are also pectinate, with tufted masses of hair behind the head, but they are narrower than the male's antennae and black in color. The mouthparts are also modified into a coiled sucking tube, which is not used for feeding.

Although Eurasian gypsy moth females are known to fly (Mikkola, 1971), the North American and some European females generally cannot fly because their large abdomens are filled with eggs. These females must climb a tree in order to lay their eggs. However, there have been several isolated observations of flight by female gypsy moths in North America (Forbush and Fernald, 1896; and Sandquist et al., 1973). It is belived that these flights were due to the females' responses to environmental factors that caused stress-induced changes in their physiology. Suggested environmental factors that may have caused these behavioral changes are overcrowding and food competition during larval development or the self-induced nervousness of the females, caused by their inability to mate as they age.

Gypsy Moth adult—male. *Courtesy: USDA*

Gypsy Moth adult—female. *Courtesy: USDA*

3

Seasonal History and Behavior of Life Stages

The time and duration of each life stage of the gypsy moth depend on the locality, weather ,and amount of food available. The gypsy moth produces one generation of offspring per year. The insect overwinters in the egg stage. Diapause or a period of suspended animation is obligatory for this insect, although some eggs may hatch prematurely. Beginning in late June, the female moth deposits egg clusters on the undersides of tree branches, on tree trunks, under loose bark, or in any shady, protected place.

Embryonic development begins as soon as the eggs are laid and is completed within four to six weeks. The fully developed larvae remain in their egg cases for seven to eight months. Exposed eggs are resistant to cold temperatures down to approximately $-7°C$ to $-4°C$, and those eggs that may be insulated beneath a snow cover can withstand considerably lower temperatures. Eggs must be exposed to cold temperatures before they hatch; however, to what temperature the eggs must be exposed is not yet known.

Of those factors influencing the beginning and end of diapause, temperature and photoperiods are the most critical. After a period of warming temperatures in late April and early May, hatching begins and extends over a period of three to four weeks. It may take more than a month for all eggs to hatch in any one locality. In cool locations, such as northern slopes, northern forests, and higher elevations, initial hatching may be delayed until mid- or late May. In Massachusetts the blooming of the shade bush is used as an indicator of the hatching of gypsy moth eggs (Marx, 1974a). Forbush and Fernald (1896) have reported the premature hatching of eggs in the fall. Although premature hatching does occur, none of the resulting early larvae can survive the winter.

Gypsy moth larvae find their food through random movement. In pure stands of favored host trees this is a simple process; but as the proportion of favored trees decreases in a stand, the problem of finding food becomes critical. The first instar ordinarily can live about one week without feeding. The normal mortality rate of early instars is high, with considerable mortality of newly hatched larvae occurring prior to feeding. Since the larvae cannot identify a host without testing the foliage, many feeding attempts may be made before a suitable host is found. Once one is found, the larvae remains on the leaves usually through the first, second, and third instars. The larvae are easily dislodged and dispersed from the host after feeding; and they may venture down from the foliage during the heat of the day to seek protection from high temperatures by crawling in bark crevices or under the ground cover.

Water becomes less of a problem as the larvae increase in size due to the decrease in evaporation surface as compared to the total body volume. The first instar larvae construct small mats of silk on the leaves for resting and molting purposes. Feeding begins about dawn, when the temperature is above $+7°C$, and continues until 10 A.M. During the afternoon they move to the undersides of leaves or other shaded places to rest and remain there during the night unless disturbed. The larvae feed voraciously and grow rapidly, devouring more and more foliage each day. For instars one through three, the caterpillars feed on the upper surfaces of the leaves by making small holes in the leaf blades. During instar three the holes in the leaves are enlarged greatly, and the caterpillars also feed from the undersides of the leaves. Larvae begin feeding on the leaf margins in considerable numbers during the latter part of the third instar.

From instars four through six, most of the feeding occurs at night between sunset and sunrise, with peak feeding occurring shortly after sunset. There are two advantages of late instar feeding at night, as suggested by Leonard (1970b): By feeding at night, the caterpillars can avoid adverse physical factors, such as high temperatures present in the daytime; and in darkness, late instars would be less exposed to parasites and predators dependent upon vision for seeking out their hosts.

Along with this dramatic change in feeding time, the larvae descend from the trees to rest in the ground cover during the daytime or seek shelter in bark crevices and under loose bark. If population densities become crowded, this feeding and resting pattern is broken, resulting in larvae feeding both day and night. The fourth and fifth instars feed voraciously, devouring approximately 24 square inches of leaves each per day. At this time larvae feed almost entirely on the margins of the leaves. Of all foliage consumed by male larvae (5 instars), 75% of the foliage is eaten in the last instar; and female larvae (6 instars) consume 65% of their total food in the last instar. Consequently, control measures against gypsy moth caterpillars should be applied as soon as possible after the eggs hatch. The caterpillars reach full size by late June or early July; after mid-July few caterpillars can be seen. The entire larval stage lasts from eight to ten weeks. Much of the biological and behavioral work concerning feeding rhythms of the larvae has been researched by Leonard (1957, 1967, 1968, 1970b, 1971, and 1972).

The development time required for each instar also has been studied by Leonard (1966). Using 12 strains of gypsy moth eggs collected in various geographic regions of the United States and Canada, he observed the following about the larvae hatching from these eggs:

1. During the first stadium, male larvae took slightly more time to complete their development than female larvae.

2. The second stadium took less time than the first, with female larvae taking longer to complete this period.

3. The third stadium was almost equal to the second in developmental time, with males again requiring more time to develop than females in this same period.

4. The fourth stadium was close to the first in duration, and the male larvae took slightly more time to develop than females.

5. For males in the fifth stadium, the time required to complete development was twice as long as that needed in the fourth sadium. The fifth stadium for females was completed in less time and was about equal in duration to the first stadium for females.

6. The last stadium for females was nearly equal in time to the fifth stadium for males.

Male larvae take from 32½ to 48 days to complete their larval development, while female larvae take from 36 to 63 days to develop. The amount of time spent in each stadium varies with temperature, light intensity, food conditions, humidity, and the population dynamics of the insect.

Male larvae may spend from 7 to 10 days in stadium one; 4½ to 6½ days in stadium two; 4½ to 6½ days in stadium three; 5½ to 8 days in stadium four; and 11 to 17 days in stadium five. Female larvae may spend 6 to 10 days in stadium one; 4 to 5½ days in stadium two; 4 to 6 days in stadium three; 5 to 6½ days in stadium four; 6 to 9 days in stadium five; and 11 to 16 days in stadium six.

After feeding, larvae can produce silk throughout each stadium, except during periods of molting and the last larval instar, when no silk is produced until the cocoon is constructed (Leonard, 1967). Throughout larval development the gypsy moth caterpillar builds resting and molting mats. These mats, composed of silk, are usually constructed wider and longer than the size of the caterpillar. The larvae show a characteristic pattern of arcing the head from side to side as the silk is laid down during the construction process. Before molting the resting mat is reinforced, and a new dense molting mat is constructed. The molting mat is used to anchor the prolegs of the caterpillar during ecdysis, allowing it to shed its exuvium. During periods of stress such as starvation the silking behavior is altered, with the larva leaving a single thread of silk during its random movements. It is also during high-population densities that early instar larvae suspend themselves from foliage on single threads of silk. Spinning larvae usually sever their silk by biting only when they come to rest upon some object, and the thread is never broken by the larvae while they are hanging freely. However, this thread may be broken by winds, which aid in the insect's dispersal.

Before pupating, the larva wanders for several hours in search of a shady, protected place. Once this place is found, the caterpillar evolves into the pupal stage in two to four days; this occurs near the end of June. The cocoon is sparse, with the prepupa and pupa wrapped in silk. The pupal stage lasts from 10 to 14 days, after which the adult moth emerges. All four stages of the gypsy moth can usually be found in one locality around July 1.

Circadian rhythm established directly by a natural diel light period, which interacts with temperature, determines the timing of eclosion, resting, communication, and courtship of the adult moth (Marx, 1973a). Temperature influences the rate at which these events (ecolosion, resting, etc.) occur and replace one another. The male moth appears four to five days before the female moth, a condition known as protandry. Both sexes do not feed, for their behavior, life pattern, and only function is reproduction.

The male moth is a fairly strong daytime flier. Duration of the moth's main flight season is normally two to three weeks, generally between mid-July and mid-August. Because the female moth is heavily laden with eggs and does not fly, she must be able to attract the male in order to ensure reproduction.

It has been known since 1888 (Hasse) that the flightless female is capable of luring males from considerable distances by means of a chemical scent. This chemical or pheromone can be released at any time from the ductless glands of the female. However, a significantly greater amount of the pheromone is released between 11:00 A.M. and 4:00 P.M. during the first three days after emergence from the cocoon than at any other time (Richerson, 1973). The average rate of pheromone release for a half-hour period is 6.5 ng, while peak emission rates may reach as high as 880 ng per half-hour period (Richerson, 1973). This burst of pheromone release occurs only once during the female's lifetime, the amount of attractant released diminishing thereafter. If the female is disturbed at any time during "calling," the ovipositor is completely withdrawn and calling ceases. Although some females have been observed to be sporadically attractive after mating (Block, 1961), the female attractancy usually disappears within the first 15 to 30 minutes after copulation begins, due to the retraction of the ovipositor and the release of masking or inhibiting chemicals over the pheromone. An inhibiting substance has been found in an inactive chromatographic fraction (Beroza, 1967), and Cardé and co-workers (1973) have identified a mating inhibitor of the gypsy moth as an olefin precursor of the sex attractant. The female does not store the pheromone.

Following the release of the pheromone, the female assumes a characteristic calling position. In this position, which may last for several hours, the female rest head upward on an upright surface with her wings slightly spread and the abdomen lowered (Doane, 1968). Once positioned, she begins a rhythmic protraction and retraction of the last segments of the abdomen. In this manner, with her ovipositor extended, the last segments of the abdomen appear cone-shaped. This appearance has been described by Snodgrass (1935) as a "provisional ovipositor." The sex glands are located on the intersegmental membrance between the eighth and ninth abdominal segments. These ductless glands lying near the surface of the tissues immediately surrounding the opening of the oviduct and copulatory pouch make the surrounding tissue attractive to the male (Collins and Potts, 1932).

The male moth flies freely upwind in a zigzag pattern to seek out the female for mating. Displaying pheromone-orientation and searching behavior by flying at slight angles to the wind, the moth responds to greatly diffused amounts of the female's pheromone. While in flight the body is held at an angle of roughly 45° to the horizontal plane, and the antennae are directly up and forward (Doane and Cardé, 1973). As the male encounters stronger gradients of the pheromone, he turns into the trail of the sex attractant and makes a direct approach to the female. The male's first response to the female is directed by pheromone detection by the antennae; and later, when within close range of the female, orientation is aided by sight, but only in the presence of the pheromone.

Once the female is located, searching behavior is terminated. The male holds his body more horizontally, and his forward flight is more direct, rapid, and usually across the path of the wind (Doane and Cardé, 1973). However, if a female is not located, the male does not persist indefinitely in searching just downwind of a pheromone source (Doane, 1968).

When two or more males compete for the same female,

they display aggressive behavior by rapidly touching each other's wings. Doane and Cardé (1973) have suggested that this behavior may be of evolutionary significance, that is, it tends to disperse males from dense to sparse populations, thereby increasing the chances of males mating with isolated females. These isolated females, having greater reproductive potential than females from dense populations, may contribute more significantly to the species when finally mated.

Mating pairs remain in copulation for at least 30 minutes (Forbush and Fernald, 1896; and Doane, 1968), although a copulation time period of only five to seven minutes is required to ensure successful insemination (Doane, 1968). Following copulation, the female shakes off the male and moves away. Displaying negative phototropism and positive geotropism, she behaves nervously while searching for a favorable spot to begin laying her eggs. An increase in the number of eggs laid and fertile eggs released occurs with increased copulation time (Doane, 1968). This may be due to the amount or concentration of sperm released over a longer period of time or to the release of associative male secretions.

Although multiple mating has been reported (Forbush and Fernald, 1896), it is considered an exceptional condition with the gypsy moth. The moth dies soon after egg laying is completed; its total life span is seven to ten days long. Most eggs are laid in July, with some being laid in late June and a few in early August. Egg masses are found almost anywhere, usually close to where the mature female larva crawled to pupate.

EGG STAGE	LARVA STAGE	PUPA STAGE	ADULT STAGE	
███				January
███				February
███				March
███				April
	███			May
	███			June
███	███	███	███	July
███		███	███	August
███				September
███				October
███				November
███				December

Life Cycle of the Gypsy Moth

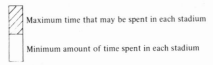
Maximum time that may be spent in each stadium

Minimum amount of time spent in each stadium

N   Larval stadium

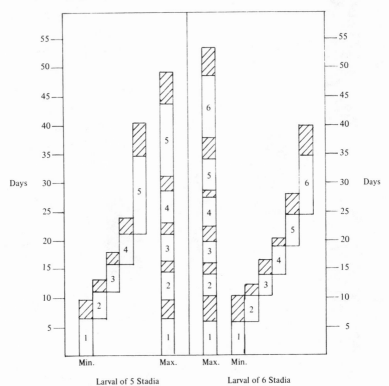

Time Spent in Each Larval Stadium

Gypsy Moth female laying eggs in a characteristic oblong egg-mass pattern. *Courtesy: USDA*

4

Dispersal

Dispersal is the movement of individuals away from one area to another, resulting in redistribution. Although dispersal is continuous within a population, it is more pronounced when population densities are high. The rate of dispersal is influenced by many factors, such as availability of food, places to live, mates, other animals in the area, man, and climatic conditions. Of all factors influencing dispersal of the gypsy moth, population pressure, wind, availability of food, and man are the most potent forces.

One-way dispersal of individuals from a population with no return is called emigration. This form of dispersal serves a basic role in population regulation and in genetic structuring of the gypsy moth. Emigration is important as a regulatory process by helping to adjust the insect's population density to some limiting resource or environmental condition. As an adaptive advantage affecting the gene pool of the population, emigration influences the quantitative and qualitative structures of the genetic make-up of the population and aids in the insect's long-term survival.

Quantitatively, the insect's genetic material would be spread more widely by dispersing individuals. This would allow more opportunities for advantageous combinations of genetic materials between emigrating individuals and those encountered in a new environment.

Qualitatively, crosses between emigrating insects and resulting offspring would exhibit greater heterozygosity and also would increase the possibility of new and advantageous genetic combinations.

Emigration thus helps to ensure long-term survival of the gypsy moth. Larvae that avoid a population crash by dispersing provide the species with a better chance of survival than those individuals which do not disperse.

In order for the gypsy moth to disperse successfully into a new area, it must satisfy at least three major requirements: It must possess the physiological capability to survive and reproduce in the new environment; it must have the ecological opportunity (unoccupied niche) to become established in the new area; and it must have physical access to the new area. All three of these requirements are met by the gypsy moth in the Northeast, where the insect disperses by both active and passive means. Of these two forms of dispersal, passive is the more serious.

Since the female gypsy moth does not fly, dispersal must be accomplished by insect stages prior to the adult. Passive dispersal is generally accomplished by newly hatched, windblown, first-instar larvae. It has been known since the early 1890s that windblown, first-instar larvae are the major vehicles of gypsy moth dispersal. These caterpillars are ideally suited for dispersal; they weigh less than one milligram when hatched and are covered with numerous long, hollow hairs, which make the larvae very light and buoyant. Under proper conditions they may be blown relatively long distances by the wind. Larval dispersal begins soon after egg hatching in May and may extend over a period of two weeks. This dispersal serves to redistribute the larvae within an individual tree, within a population community, or to a new area.

Before the larvae begin feeding or when disturbed during

high population densities, they arch their bodies and string downward from the foliage on silken threads. It is during this particular behavior pattern that many are swept up by the wind and convectional currents and dispersed. The strands of silk are broken, and the larvae are carried away, with silk thread still attached. Larvae may be extended perpendicularly to the tree trunk by wind currents as slow as 1.3 miles per hour (Marx, 1970b). The distance they are carried by air currents depends somewhat on the length of silk attached to their bodies.

Larvae have been captured more than 35 miles from the nearest known infestations (Collins, 1917; and Nichols, 1962). Some larvae have been captured on screens attached to an airplane at height of over 2,000 feet (Collins and Baker, 1934). Once larvae reach such heights, they may be blown by cross currents opposite to the surface winds. Larvae have been trapped at wind velocities as low as two miles per hour, but a substantial increase in the number of dispersed larvae occurs at velocities of eight miles per hour (Collins, 1915).

Newly hatched larvae spend 12 hours or more on the egg mass before moving to the nearby foliage. Being photo-positive, they ascend the trees and move to the highest points of the tree crown or to the side closest to the sun. This ascent is accomplished in order to disperse, not to feed. Behavior patterns of the young larvae are directly affected by environmental conditions. If cold, wet weather occurs after hatching, the larvae usually stay around the egg mass until conditions are more favorable for foraging. The larvae are active only when the temperature is about $+18°C$, the larvae are dark in color; by absorbing solar radiation, they may initiate activity at lower temperatures.

Dispersal by early-instar larvae occurs throughout stadium one, primarily on hot, dry days ($+18°C$ to $+24°C$) threshold activity temperature for larvae. However, the when larvae are active, and convectional currents from the heated ground surface are prevalent, along with cumulus clouds. Peak dispersal is usually in the morning hours, coinciding with the active behavior patterns of the larvae, and occurs about one week after the peak of hatching. In dense infestations of early instars, many do not feed but arch their bodies and release their attachments to the substrate, which makes them more easily dispersed by the wind.

Larval dispersal decreases on cold or rainy days when the larvae are relatively inactive. On these days larvae are usually on the undersides of leaves and branches or in bark crevices and are not easily dislodged by the wind.

Since its introduction into North America, the early spread of the gypsy moth had been to the north and northeast, coinciding with the general direction of the prevailing surface winds after the hatching season. Dispersal northward by the insect may be limited through egg mortality by freezing temperatures. The gradual movement of the gypsy moth northward from its initial point of introduction in Massachusetts has not been accompanied by natural selection for a more cold-hardy race (Sullivan and Wallace, 1972). The inability of the gypsy moth to develop a more cold-hardy race in North America may eventually halt the dispersal of the insect in that general direction. At the insect's present level of cold-hardiness, it can be expected to increase its range northward and westward to include much of the area bounded by the $-34°C$ isotherm throughout southern Quebec and Ontario (Sullivan and Wallace, 1972). In addition to the present cold-hardiness of the eggs, any protection given the egg mass by snow cover may be of considerable importance to the survival of the insect and its increased range in North America.

Due to the current inability to limit population increases and spread of the gypsy moth in Pennsylvania, New York, and New Jersey, large numbers of gypsy moths will gradually disperse southward by passive and active means.

When dispersal occurs on a large scale, woodlands several miles away may experience defoliation even though no previous infestations had existed. Larval dispersal by the wind provides the species with a gradual enlargement of its range. Therefore, serious threats to woodlands within 20 to 40 miles of gypsy moth infestations exist, but woodlands further removed are not immediately threatened by the insect. This is due to the relatively large distance of separation between noninfested and infested woodlands. The greater the distance the larvae are carried by the wind, the greater the distance between them and, therefore, the smaller the chance to fall closely enough together to establish a significant community capable of causing considerable damage. In addition, in order for an infestation to occur and exist over a prolonged period of time, there must be a large number of larvae present to offset natural mortality. If it were not for the high mortality rate of the larvae, the gypsy moth might have spread further to become esablished in more localities than it already has.

If gypsy moth populations do develop primarily as a result of wind-borne sources, they may do far more damage to woodlands than gypsy moth populations on the periphery of an established infestation. This is due to the immediate lack of parasites and predators in new wind-borne infestations and a resulting unchecked rate of increase in numbers of the wind-borne population. Once a collapse occurs in an established infestation, large numbers of parasites move out of the infestation center into the adjacent woodland, where they parasitize gypsy moths in the periphery of the collapsed infestation. Often wind-borne infestations are too far removed from an established or collapsed infestation for parasites and predators to reach.

The generally infested areas of the Northeast are enlarged each year as a result of millions of larvae that are blown from spot infestations and fall nearby, spreading the range of the gypsy moth a short distance. These wind-borne larvae succeed in gaining a foothold, and the population expands. Once these dispersing individuals have established a new infestation, they serve as source points for further expansion. This gradual enlargement of the infested Northeast is expected to continue southwestward, down the ridges and valleys of Pennsylvania and West Virginia. Since the prevailing winds in the Northeast are to the north and east, the rate of spread of the gypsy moth to the southwest is not dramatic.

Large caterpillars disperse actively by crawling; however, the distance of their dispersal movement is not great. Emigration of large caterpillars is enhanced when population densities are high and when the foliage present is not to their liking. On some occasions when the larvae begin to disperse, they form a trail of silk to make themselves more readily dispersible by the wind. Dispersal of large caterpillars out of defoliated woodlands usually occurs from mid- to late June. During heavy infestations when food is scarce, caterpillars commonly crawl from one tree to another in search of food. This wandering of large numbers of larvae in search of food and sudden outbreaks of the insect have focused an unusual amount of attention on the gypsy moth.

Dispersal of members of a population and their resulting infestation of a new area involve a number of parameters. These include the mode of dispersal (active or passive), the frequency of dispersal (number of disseminules per unit area), the physiological condition of the dispersing individuals, and the condition of the new environment. McManus (1973) has identified the environmental parameters affecting the dispersal process of the gypsy moth. The two most important factors influencing major dispersal or range of the insect are the forest type invaded and the mortality rate caused by low winter temperatures.

One form of passive dispersal is artificial dispersal, the

principal means of this dispersal being man's transfer of egg masses, pupae, or larvae on the bark of trees and forest products, on stones and quarry products, on scrap iron, or on such objects as trailers, tents, campers, and so on. This form of movement often results in long-range transfer of the gypsy moth. This problem occurs when infested vehicles or materials are moved into uninfested areas to the south and west of the generally infested Northeast. Although quarantines are in effect, artificial dispersal is a serious threat to woodland communities far removed from the major infested areas.

The dispersal of the gypsy moth hinders any management or control program against the insect. From areas of dense or sparse population levels, larvae are blown or accidentally transported to new localities, where they would not normally occur in troublesome numbers and remain unnoticed for many years. This results in population increase, yielding more widespread epidemics of the gypsy moth.

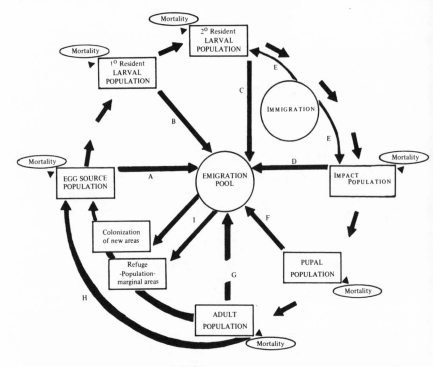

Dispersal of the Gypsy Moth

From generation to generation, regardless of mortality rates experienced by the insect's life stages, the gypsy moth may emigrate from its forest community. The emigrating individuals may collectively be referred to as an emigration pool. Eggs (A) may be removed from an indigenous population by passive dispersal means to the emigration pool. The primary residents (B) of the larval population may disperse to the emigration pool by passive, wind-borne dispersal. The secondary residents (C) of the population may contribute to the emigration pool by both active and passive means. These secondary residents of the population give rise to the impact population, which may contribute slightly to the emigration pool (D). It is during the larval stage of development that gypsy moth individuals from a non-indigenous source may immigrate to the native forest community and create both an additional source of impact individuals and crowded conditions conducive to further emigration (E). Resulting pupae from the impact population may contribute little if any to the emigration pool and only in a passive manner (F). Adults (G) are insignificant when considered as an emigration means for the population, but they do serve to perpetuate the indigenous population with egg masses, and possibly to increase the number of individuals and size of infested area (H). It is the emigration pool that establishes new areas of infestation and marginal infestations that eventually merge with the indigenous population, creating conditions suitable for the perpetual dispersal of more individuals (I).

The nuisance problem created by migrating caterpillars is shown
in this photograph. *Courtesy: USDA*

5

Gypsy Moth Populations and Density Dependent and Independent Factors

Although the destructive stage of the gypsy moth is the larval stage, the pest population is composed of groups of gypsy moth individuals in all stages of development in a particular habitat. Populations of the insect rise and fall in response to weather conditions, disease, parasites, predators, and starvation, but they seem to be noticed only when abundant. It is at such high population levels that they cause visible defoliation.

Generally, for the first three to five years following the gypsy moth's infestation of a new area, the insect remains unnoticed while distributing itself at low densities throughout the area. If little parasitism and predation occur in the new area, accompanied by large egg masses and high survival rates for young larvae, the gypsy moth can go from insignificant numbers causing light damage to massive numbers causing complete defoliation.

Gypsy moth numbers vary greatly from season to season and from region to region. When certain critical population density levels ("blow-out levels") are reached, there is a

71

change in larval behavior, resulting in an increase in wind dispersion. This may result in sudden population density changes. After two or three years of heavy defoliation, populations usually decline or crash the following year. This same syndrone of abrupt population change has occurred among other native woodland pests, such as the cankerworm or inchworm, *Alsophila pometaria,* the orange-striped oak worm, *Anisota senatoria,* and the forest tent caterpillar, *Malacosome disstria.* However, the variation in the patterns of noticeable feeding and massive defoliation is not cyclic with the gypsy moth in terms of regularity of outbreaks.

Forbush and Fernald (1896) reported an average growth rate for the gypsy moth of 6.4-fold from generation to generation, but it is not uncommon for an average growth increase to be 7.5- or 8.0-fold. Growth increases usually are higher in low-density populations than in high-density populations, largely due to density-dependent suppressive forces. Highest growth increases and resulting destructive populations occur in those areas where oaks occupy 15% to 25% or more of the forest stand.

For purposes of pest control and management, it is practical to assume a higher rate of population growth for the gypsy moth than those rates normally experienced in the field. Because of this the insect is considered to increase at a 10-fold rate per generation. With this rate of increase per generation or year, the trend of an uncontrolled gypsy moth population would increase on a geometric scale, and its resulting populations could cause damage on a geometric scale. Control of gypsy moth populations is difficult because young populations are highly unstable and show a continual redistribution of individuals throughout a relatively large area over a short period of time.

Campbell (1967) has presented an analysis of numerical changes in gypsy moth populations. This analysis is based upon the identification of age-interval survival rates and

mortality-causing agents related to these rates, the construction of mathematical models for each mortality-causing
agent, and an overall investigation of long-term trends in
gypsy moth numbers.

The biology and ecology of gypsy moth populations are
controlled to varying degrees by density-dependent and
density-independent factors. Density dependent includes
those forces caused by the insect population itself, while
density independent are those environmental elements that
exert a direct or an indirect effect upon the insect population,
regardless of the population size or structure.

As the population size of the insect community fluctuates
over a period of time, the corresponding increases and
decreases in numbers of individuals apply direct pressure
on the individuals, causing changes in insect behavioral
patterns, developmental rates for larvae and pupae, fecundity and mortality rates, dispersal patterns, and insect
longevity.

Environmental elements that influence population size
or insect behavior are temperature, rain and moisture,
evaporation, light, and geographic location.

DENSITY—DEPENDENT FACTORS

Population density has been shown to influence insect
behavior, morphology, and physiology (Uvarov, 1921).
Leonard (1970a) has presented evidence to suggest that the
gypsy moth is numerically self-regulating through a shift
in the quality of individuals induced by changes in nutrition.
Under crowded conditions and during scarcity of food,
larvae undergo additional molts, and the adult gypsy moths
emerge sooner in dense populations than in sparse ones
(Leonard, 1968). In North America up to eight instars or
seven molts for males and nine instars or eight molts for
females have occurred under crowded conditions (Leonard,

1970a). The significance of this polymorphism and any change in egg structure or content appears in the roles they play in such density-regulating factors as dispersal, behavior, developmental rates, and fecundity.

Eggs having reduced nutrient reserves produce larvae that undergo additional molts. These eggs, from moths whose larval stage was subjected to crowded conditions, produce a new larval population with a prolonged stadium one and an increased number of molts. Larvae with prolonged stadium-one periods are more readily dispersed by the wind (Leonard, 1970a). Because these larvae are rendered dispersible for a longer period of time, the gypsy moth can regulate to some degree its population numbers in a given locality. In addition, under crowded periods many larvae are forced to disperse before feeding because no food is available. Most airborne larvae appear to be unfed; or, if feeding has occurred, the amount of food eaten is considerably less than the amount eaten by larvae on leaves (Leonard, 1970a).

Diel periodicity of late-instar larval feeding patterns is interrupted during overpopulated periods of time. Late-instar larvae normally feed at night and rest during the day; however, under crowded conditions larvae are active both night and day competing for food and resting sites.

During early-instar larval development, when large numbers of larvae are competing for the same hosts, most larvae do not feed because of continuous interruption from other larvae searching for food; the interrupted larvae become nervous and agitated. If nervous enough, the slightest disturbance will cause these caterpillars to drop from the foliage on silken threads. From here they may be blown away, reducing competition for food and thereby enabling the remaining larvae to feed without interruption (Campbell, 1970).

The only phase of gypsy moth development prolonged

by crowded conditions is the first stadium. Crowding results in a supernumerary condition, indicating that a qualitative change has occurred, and results in a reduction of pupal weight and fecundity. Adults emerging from smaller-than-average pupae produce fewer eggs. The male larvae that hatch from these eggs may appear lighter in coloration, causing the phenomenon of phase polymorphism (Leonard, 1974). Male moths whose developmental stages have experienced crowded populations also may appear lighter in coloration (Leonard, 1968).

The most important factor affecting gypsy moth population dynamics under crowded conditions is the increase of natural dispersal.

DENSITY—INDEPENDENT FACTORS

Temperature

Climatic factors have a profound effect upon gypsy moth populations. Some egg masses are killed when exposed to a temperature of $-26°C$, while almost total mortality occurs at $-32°C$ (Summers, 1922). Duration of cold at lower ranges of lethal temperature is as important as the temperature itself. Eggs exposed to $-25°C$ for one day averaged 21% mortality; after two days exposure, 61% mortality occurred; after three days 93% mortality was observed; and after five days, 96% mortality was reported (Maksimovic, 1958). Minimum winter temperatures in northern Maine, New Hampshire, and Vermont are low enough to kill eggs not protected by snow insulation. In such cold locations, selection may favor those insects which lay their eggs close to the ground to acquire snow insulation from winter temperatures (Leonard, 1972). It is in such cold regions as northern New England that the gypsy moth has become slowly established. Cold winter temperatures, therefore, may be considered as a limiting factor against the northern spread of the insect in North America.

Temperature also has a striking effect on larval development. Increasing temperature accelerates the developmental rate of larval stages. Maksimovic (1958) illustrated that male and female larvae complete their development in 25 to 27 days respectively at $+32°C$, while they require 92 to 97 days for development at $+22°C$.

Although a cold-hardy race of the gypsy moth has not developed in North America, selection of such a race appears to be developing in some parts of Europe and Asia (Pantyukhov, 1964; and Jankovic et al., 1959). Pantyukhov (1964) demonstrated that eggs from eastern Russia could withstand a temperature of $-44°C$ for short periods of time.

Eggs are normally protected from winter temperatures by snow insulation, but they also may be protected from these cold temperatures by ice or by the covering of setae deposited on the egg masses from the abdomen of the female. However, the insulation provided by the female setae is not so great as that provided by the snow or ice. Summers (1922) noted the importance of snow-insulating properties for the gypsy moth eggs and for their eventual hatching. He also observed that snow drifting around tree trunks causes some eggs to be protected and others to be exposed.

Rain and Moisture

During periods of rainy weather and low temperature, considerable mortality of newly hatched larvae occurs. Late frosts in spring may indirectly sustain heavy mortality rates in larvae as a result of starvation due to severe damage to woodland foliage.

Rain and moisture affect larvae in many ways. During periods of rainfall, dispersal is reduced since larvae remain attached to the undersides of leaves or in bark crevices;

at such resting spots they are not easily dislodged by winds (Leonard, 1971). High humidity is often accompanied by increased mortality rates. This is credited to the increased incidence of a nuclear polyhedrosis virus during periods of relatively high humidity (Wallis, 1957 and 1962). As relative humidity increases, there is a general trend toward a reduction of the time period required for hatching and an increase in longevity for adult moths.

Evaporation

Evaporation is also known to affect larval behavior. As the evaporation rate increases, feeding activity also noticeably increases. The loss of water from the caterpillar's body is considered as a stimulus directly affecting feeding behavior. The effects of evaporation may explain the more rapid feeding in the upper parts of the forest canopy than in the lower parts. As larval feeding opens the crown canopy, feeding activity increases in the lower parts of the canopy with the increase of light penetration, temperature, and air circulation due to the gradual opening of the canopy.

Light

Light also affects larval behavior. Young larvae are attracted and older larvae repelled by strong light intensities. Newly hatched caterpillars are positively phototactic and climb upward to the foliage after hatching. This response can be inhibited if the tactile stimuli to the forelegs are lost, as would be the case at the end of a branch (Zanforlin, 1970). Diel periodically feeding rhythms are controlled by light, with early-instar larvae feeding shortly after sunrise and late-instar larvae feeding just before sunset.

Geographic Location

Geographic variations and corresponding climatic conditions show related variations in the developmental rates of larvae in each geographic region. For example, larvae from southern Canada are known to develop faster than those from Connecticut (Leonard, 1966).

Forest site conditions often determine whether larvae remain on the tree or seek resting sites in the ground cover (Bess, 1961; and Bess et al., 1947). Since larvae usually pupate in the same site utilized as a resting place during the day, forest site conditions, along with larval and adult behavior, determine the height at which egg masses are laid. This also somewhat determines whether the eggs will die of winter temperatures, if they are laid too high to be protected by snow insulation. In mesophytic hardwood stands larvae seek shelter in the ground cover, and the eggs are laid close to the ground, being protected by snow insulation during the winter (Leonard, 1972).

Table 5 - 3

TREND OF AN UNCONTROLLED GYPSY MOTH POPULATION
INCREASING AT A 10-FOLD RATE PER GENERATION

Generation or Successive Years	Number of Adults in Population
1	2
2	20
3	200
4	2,000
5	20,000
6	200,000
7	2,000,000

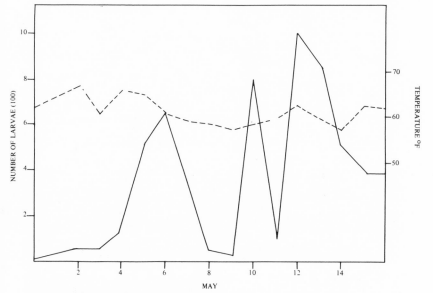

Effect of Rain on Larval Dispersal

On May 8, 9, and 11, days of rainfall, the number of larvae trapped on nets was reduced because of reduced larval dispersal. The dotted line indicates the temperature at 10 a.m. each day. *Courtesy: Entomological Society of America, after Leonard, 1971*

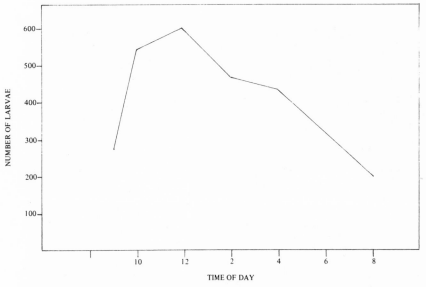

Time of Day of Early-Instar Dispersal

Average number of dispersing larvae trapped between 8 a.m. and 8 p.m. on May 4, 6, 12, 14, and 15. *Courtesy: Entomological Society of America, after Leonard, 1971*

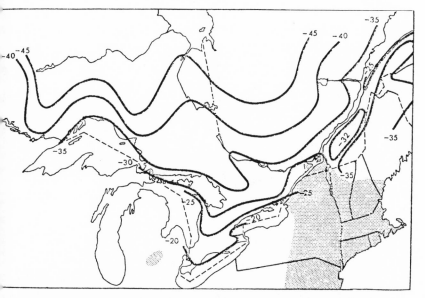

Future Distribution of the Gypsy Moth to the North and West.
*Courtesy: The Canadian Entomologist, after Sullivan and Wallace,
1972.*

The potential spread of the gypsy moth northward and westward into Canada is expected to include much of the area bound
by the −30 degree isotherm. Stippled area is present distribution
of the gypsy moth.

6

Plant Hosts and Damage

Unlike most woodland defoliators that restrict their feeding to a single species or group of plants, the gypsy moth is a general feeder on the leaves of forest, shade, and fruit trees and shrubs. In North America a total of 485 plant species has been recorded as food sources for the gypsy moth (Forbush and Fernald, 1896). It is due to this wide range and variety of host plants, over 150 primary hosts, that the gypsy moth is considered such a serious woodland pest. It is also the only significant insect in the northeastern United States that feeds on both hardwoods and conifers. In heavy infestations few species of trees are overlooked; and when food is scarce, grasses, vines. flowers, and garden and field crops are sometimes eaten. Two or more defoliations can kill some hardwoods, while one complete defoliation may kill some evergreen trees. Pines are more resistant to defoliation than hemlocks, but severe defoliation also kills them, and they may die two or three years after being defoliated. Oaks usually die if they are defoliated in successive years with 60% of the tree foliage removed each year. In contrast, larch withstands repeated defoliation for ten years or more. However, in most cases, one year of defoliation weakens the trees and makes them vulnerable to secondary attack by other insect pests or diseases.

The total acreage severely defoliated by the gypsy moth fluctuates greatly from year to year. During a three-year period (1953–1955) in New England, over two million acres were severely defoliated; in 1953, one and a half million acres; in 1954, 491,000 acres; and 52,000 acres in 1955 were defoliated. Since a single, mature caterpillar can eat a square foot of leaves per day, these high defoliation figures are easily realized by moderate to heavy gypsy moth infestations.

There are distinct differences between young and old larvae for food or host preferences. In early instars, the larvae primarily feed on such host spcies as oaks, *Quercus*; but as the larvae mature, elm, *Ulmus*, willow, *Salix*, poplar, *Populus*, and pine, *Pinus*, are readily fed upon. The differences in host preferences for young and old larvae can be categorized in four basic groups. Group 1, dominated by the oaks, includes those species of plants favored by all instars. Group 1 is composed mainly of hardwoods, and these trees must be present for younger larvae to feed initially. Trees favored by the gypsy moth in Group 1 that also are of commercial value are white, red, and black, chestnut, scarlet oak, American and big-tooth aspen, and paper birch. Of the species in Groups 2, 3, and 4, either the larvae cannot feed upon them in their early instars, or, if they are forced to feed upon them, few survive.

Group 2, dominated by pines and spruces, includes those species of plants favored by the older instars. A single defoliation is often capable of killing many of the evergreens in this group.

Group 3 plants are sometimes defoliated when a dense population increase occurs. These host plants are moderately favored by older instars and allow a small proportion of the larvae to develop. Cherry, hickory, and maple trees dominate this group. Trees of commercial value in Group 3 on which the larvae cannot feed in early-instar stages are red, silver, sugar, and Norway maples.

Through feeding studies, combinations of various plants in Groups 1, 2, and 3 have been tested for their ability to attract the gypsy moth larvae (Doskotch, 1973). There may be several chemical feeding stimulants present in those plants which encourage the larvae to feed upon them.

Group 4 is composed of those species not normally favored by any larval instar; however, under stress conditions such as starvation, noticeable feeding may occur on these species. Of all the trees in this group, ash, *Fraxinus,* is perhaps the most naturally undesired or immune of any broad-leaved tree to gypsy moth feeding. Research on the chemical nature of the leaves, which causes this group to be unfavored by the gypsy moth, produced the speculation that Group 4 plants may be divided into two smaller categories.

One category would include such trees as *Aesculus glabra, Populus deltoides,* and *Platanus occidentalis,* all of which lack feeding stimulants but are acceptable when mixed with a plant extract that is a stimulant. A second category, comprising trees such as *Liriodendron tulipifera* and *Catalpa speciosa,* which all contain deterrents to larval feeding, would be the more resistant to attack. These plants can counteract the chemical stimulant present in the oaks.

Through the study of these chemical stimulants and deterrents, it is hoped that a material can be synthesized to prevent larvae from feeding on favored foliage. Although such trees as oaks do have stimulants for feeding and ash trees have deterrents to feeding, many trees contain neither chemical compound. However, these trees that are neither favored nor unfavored (cottonwood and sycamore, e.g.) do contain a synergistic compound when in combination with favored foliage.

The amount of damage or proportion of foliage removed

by the gypsy moth in any place and year depends not only upon the insect's population density but also upon a number of other factors, including climatic conditions, forest site, and forest composition. Thus a multitude of defoliation levels may result from a particular population size. Forest composition is a significant factor influencing the activity of the gypsy moth and the most important single factor affecting its intensity of infestation and population dynamics. In forest stands where the percentage of favorable hosts is high, the danger of severe infestation is also high, while the danger of infestation is almost nonexistent in pure stands of unfavorable hosts. The manner in which distribution and degree of damage coincide with forest regions and compositions is striking and significant (Behre et al., 1936).

There are four major forest regions in the northeastern United States: the oak-pine region of Cape Cod and southeastern Massachusetts; the white pine region in eastern Massachusetts, southeastern New Hampshire, and southwestern Maine; the northern hardwood region of western Massachusetts, central Vermont, northern New Hampshire, and northeastern New York; and the chestnut oak-yellow poplar region of Connecticut, Rhode Island, southeastern New York, and parts of New Jersey and Pennsylvania.

Heavy defoliation has occurred throughout the oak-pine region where the forest composition is predominantly oak, which provides ideal conditions for gypsy moth feeding and infestation. In contrast to this region, there has been no serious defoliation in the northern hardwood area where unsuitable host trees such as maple, yellow birch, and beech predominate. In the white pine region of Massachusetts where the forest composition is mainly mixed hardwoods with some stands of pine, repeated heavy defoliations have occurred. Little defoliation has been observed in the oak-yellow poplar region of the Northeast due to the low percentage of favorable host species. Unfavorable hosts such as chestnut, yellow poplar, maple, hickory, black and yellow birch, and ash are very abundant in this area. However, when the percentage of oaks in a forest stand increases, as in many parts of the oak forest in Pennsylvania and New Jersey, heavy defoliation has occurred. Forest researchers continue to study the ecological relationships between forest site and gypsy moth susceptibility or resistance and the effects of defoliation.

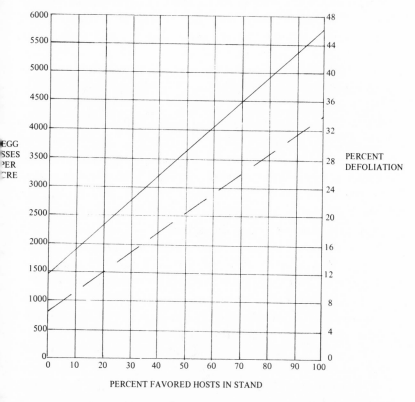

Relations between Proportion of Favored Hosts in Stands and Intensity of Gypsy Moth Infestation and Resulting Defoliation. *Courtesy: Massachusetts Forest and Park Association, after Behre et al., 1936*

The accompanying graph shows the correlation between percent of defoliation, percent of favored trees in a stand, and number of egg clusters per acre. As illustrated in the graph, as the proportion of favored trees increases, there is a corresponding increase in the amount of infestation to almost 6,000 egg clusters per acre.

The ability of the gypsy moth to establish itself in any locality directly correlates with the percentage of favorable host species present in the locality. If the percentage of favorable hosts is less than 20%, the danger of gypsy moth infestation may be considered minimal. Generally, the degree of infestation and defoliation positively correlates with the percentage of oaks in a stand. However, low insect mortality rates and heavy larval migration may result in dense egg mass populations and heavy defoliations in areas where favored hosts are low in percentage composition of the stand.

Bitzer (1971) determined the effects of three years of successive defoliation on an 880-acre oak ridge in southwestern Monroe County, Pennsylvania. His analysis indicated a very significant percentage of oak mortality and a distinct host preference for the oak species, with a corresponding loss of timber stumpage. Following the elimination of oak species due to gypsy moth defoliation, a stand of red maple, black gum, and sweet birch, all unfavored species of the gypsy moth, began to dominate the area. The damage incurred on this ridge was as follows: (1) 58% of the oak trees were dead by the third year of defoliation, followed by complete destruction of all oak trees in ensuing years; (2) none of the non-oak species died from gypsy moth defoliation after three years; and (3) 83% of the sawtimber value and 64% of the pulpwood in this stand was in dead or declining trees (Bitzer, 1971).

Another example of the extent of damage caused by gypsy moth defoliation can be seen in the Newark, New

Jersey, Watershed. Infestations were first observed in the Watershed in 1968, and over the next three summers more than 17,000 acres of woodlands sustained repeated defoliation. During this same period more than 1,000,000 oaks, 39,000 eastern hemlock trees, and 8,000 white pine trees were killed as a direct result of gypsy moth attack (USDA, 1973d).

In 1969 over 10,000 oak trees in the Morristown National Historical Park in New Jersey died from persistent gypsy moth defoliation (Marx, 1970a). In 1970, 100,000 acres in New York and an additional 100,000 acres in New Jersey were defoliated to various degrees by the gypsy moth (Marx, 1970a).

A recent example of gypsy moth damage and selective feeding was observed on 15,880 acres in Pike and Monroe Counties, Pennsylvania, over the two-year period of 1971 and 1972 (Quimby, 1972). Oaks were fed upon selectively by gypsy moth larvae and received the greatest part of the defoliation impact. White, red, and chestnut oaks were heavily defoliated while birch, hemlock, and other trees sustained negligible damage. The oaks, which constituted 88% of the total sawtimber volume, represented 16% dead sawtimber value and 32% declining sawtimber value for this stand (Quimby, 1972). Of all the trees (six inches or more in diameter) in this area, 900,000 or 57% were dead from gypsy moth defoliation (Quimby, 1972).

The health of a forest stand can affect the activity and mortality rates of early-instar larvae, which are sensitive to food quality. Not only do the leaves of healthy stands of favored hosts yield nourishment to early instars and aid in reducing mortality rates, but also the leaves of trees weakened once by defoliation or drought provide additional nourishment. In spring, leaves normally contain high concentration of oils, sugars, and starches; however, those trees weakened by defoliation or drought have increased concentrations of sugars, starches, and foliar proteins (Amirk-hanova, 1962). The leaves of trees weakened repeatedly by defoliation provide little in terms of nutrition.

The degree to which the forest canopy is open also may affect the population density of egg masses and be a critical factor in the rapid increase of population numbers. Golubev and Semevsky (1969) demonstrated the population density of gypsy moth egg masses to be 44 times greater in disturbed forest environments such as forest borders and forest roadsides than in deep forest regions. Bess and co-workers (1947) noted that the densities of gypsy moth infestations increased at the edges of woodlands and inhabited regions.

Gypsy moth attacks and the resulting degree of destruction to trees, regardless of their growth stage, in a forest stand influences the rate of growth, tree form, species composition, and ecological succession of the stand. Damages caused by the gypsy moth are either the direct or indirect results of primary attack.

Direct losses to attack are many. Killing or weakening of trees is obvious, but weakening forest trees is a more serious threat than killing them, for thousands of trees are usually weakened for every one killed. Weakened trees are susceptible to secondary injuries caused by such agents as fungi and other insects. Retardation of woody growth, wood destruction, and decreased seed production also accompany gypsy moth attack. Most conifers, especially white pine and hemlock, are killed by a single, complete defoliation while many hardwoods are killed by two successive defoliations. Food reserves in trees may become critical after two years of defoliation, and the death of trees is usually the result of heavy, sustained feeding.

The stand composition of a woodland may be altered by gypsy moths that selectively attack favorable host plants, which eventually become reduced in numbers or eliminated from the stand. Whether this change in stand composition is for better or worse depends on the value of the trees removed and of those that remain.

Indirect losses from gypsy moth defoliation include increased fire hazards, reduced recreational values, damage to the aesthetic value of wodlands, and rendering woodlands vulnerable to the transmission and inoculation of disease-producing agents. Tree growth aids in the retention of ground moisture and prevents erosion. A defoliated area may become more susceptible to fire by the presence of large amounts of dead timber and an accelerated rate of moisture evaporation from the soil due to the loss of tree growth and to an open canopy during the most critical growth period of the summer. In addition, the water supplies of many small streams used in municipal or commercial reservoir systems are also diminished with increased evaporation of soil water. The recreational and aesthetic value of a woodland may be reduced by unsightly dead timber and migrating larvae, which are a nuisance to many.

Tree defoliation also has far-reaching effects. Mortality rates of nestling birds are increased with greater evaporation and heat due to the openness of the forest canopy (Cobb, 1968). High ground temperatures due to the loss of foliage in defoliated areas cause snakes to migrate to lower altitudes; even wildlife suffers from reduced vegetation, especially the decrease in acorn production.

In terms of economic value lost through gypsy moth defoliation, it is very difficult to measure the full impact. Depending on the forest composition and stand site, estimated loss of seven to ten dollars per acre of defoliated commercial forests in the Northeast is not uncommon. Losses in dollars when recreational areas are affected far exceed losses in timber or stumpage value.

The greatest damage from the gypsy moth is sustained in the oak forests and on ridges and plateau areas that support vast stands of white and chestnut oak. These hosts and locations are the favorite breeding grounds for the insect. Oak stands represent ideal habitats for the insect and since 1950 have undergone large-scale mortality in many parts of Pennsylvania, New Jersey, and New York. Large outbreaks of the gypsy moth in New Jersey have occurred in the Kittatiny Mountains from Lake Success to Catfish Pond, in Bowling Green, Campgan, Copperas, Green Pond, Hamburg, Pochuck, Ramapo, and Sparta Mountains, and in Cape May, Monmouth, and Ocean Counties. In Pennsylvania large outbreaks have occurred in Berks, Carbon, Centre, Dauphin, Lackawanna, Lebanon, Luzerne, Mifflin, Monroe, Schuylkill, and Union Counties.

In direct contrast to these oak stands is the northern hardwood forest, which occupies parts of New Jersey, New York and Pennsylvania. This forest is composed primarily of beech, birch, sugar maple, ash, and black cherry and has experienced few serious insect problems It is considered by many to be the most insect-resistant forest type in North America.

SPECIES THAT ARE FAVORED FOOD FOR ALL GYPSY MOTH LARVAL INSTARS

GROUP 1

Common Name	Scientific Name	Common Name	Scientific Name
Alder, speckled	*Almus rugosa*	Oak	*Quercus*
Apple, common	*Malus pumila*	beer	*Q. ilicifolia*
Aspen	*Populus*	black	*Q. velutina*
balsam popular	*P. balsamifera*	black jack	*Q. marilandica*
American	*P. tremuloides*	bur	*Q. macrocarpa*
big-tooth	*P. grandidentata*	chestnut	*Q. prinus*
Lombardy popular	*P. nigra*	chestnut, dwarf	*Q. muhlenbergii*
Balm-of-Gilead	*P. candicans*	chestnut, rock	*Q. prinoides*
Birch	*Betula*	pin	*Q. palustris*
gray	*B. populifolia*	post	*Q. stellate*
paper	*B. papyrifera*	scarlet	*Q. coccinea*
river	*B. nigra*	shingle	*Q. imbricaria*
sweet	*B. lenta*	red, northern	*Q. rubra*
Blueberry	*Vaccinium angustifolium*	white	*Q. alba*
Boxelder	*Acer negundo*	white, swamp	*Q. bicolor*
Gum, sweet	*Liquidambar styraciflua*	Rose, prairie	*Rosa setigera*
Hawthorn	*Crataegus*	Serviceberry	*Amelanchier*
cockspur	*C. crus-galli*	Allegheny	*A. laevis*
frosted	*C. pruinosa*	downy	*A. arborea*
Hazelnut	*Corylus*	Spruce, blue	*Picea pungens*
beaked	*C. rostrata*	Sumac	*Rhus*
common	*C. americana*	dogwood	*R. vernix*
Larch	*Larix*	dwarf	*R. copallina*
American	*L. americana*	smooth	*R. glabra*
European	*L. decidua*	staghorn	*R. typhina*
mountain	*L. laricina*	Willow	*Salix*
Linden (Basswood)	*Tilia*	glaucous	*S. discolor*
American	*T. americana*	sandbar	*S. interia*
European	*T. cordata*	white	*S. alba*
Mountain-Ash, American	*Sorbus americana*	Witch-hazel	*Hammelis virginiana*

Table 6-7

SPECIES THAT ARE NOT FAVORED FOOD FOR ALL GYPSY MOTH LARVAL INSTARS

GROUP 4

Common Name	Scientific Name	Common Name	Scientific Name
Arborvitae	*Thuja orientalis*	Locust, black	*Robinia pseudoacacia*
Arrowwood	*Fraxinus*	Maple	*Acer*
Ash		mountain	*A. spicatum*
black	*F. nigra*	striped	*A. pennsylvanicum*
blue	*F. quadrangulata*	Mountain-laurel	*Kalmia latifolia*
green	*F. pennsylvanica*	Mulberry	*Morus*
	var. lanceolata	red	*M. rubra*
pumpkin	*F. profunda*	white	*M. alba*
red	*F. pennsylvanica*	Osage-orange	*Maclura pomifera*
white	*F. americana*	Osier, red	*Salix viminalis*
Azalea	*Rhododendron*	Pepperbush	*Clethra alnifolia*
flame	*R. calendulaceum*	Persimmon, common	*Diospyros virginiana*
white	*R. viscosum*	Poison-ivy	*Rhus toxicondendron*
Balsam, fir	*Abies balsamea*	Privet	*Ligustrum vulgare*
Baldcypress	*Taxodium distichum*	Raspberry	*Rubus sp.*
Catalpa, southern	*Catalpa bignonioides*	Sarsaparilla	*Aralia nudicaulis*
Cornus (Dogwood)	*Cornus florida*	Skunkcabbage	*Symplocarpus foetidus*
Cypress, bald	*Taxodium distichum*	Sweetbrier	*Rosa rubiginosa*
Dangleberry	*Gaylussacie frondosa*	Sycamore, American	*Platanus occidentalis*
Dock, narrow	*Rumex sp.*	Tulip poplar (Tuliptrea)	*Liriodendron tulipifera*
Elder, American (box)	*Acer negundo*	Viburnum, sweet	*Viburnum lentago*
Feverbush	*Benzoin aestivale*	Virginia creeper	*Psedera quinquefolia*
Grape	*Vitis sp.*	Walnut, black	*Juglans nigra*
Greenbrier	*Smilax rotundifolia*	Willow, bay-leaved	*Salix pentandra*
Hardhack	*Smilax tomentosa*	Winterberry, smooth	*Ilex laevigata*
Holly	*Ilex opaca*		
Honeylocust	*Gleditsia triacanthos*		
Honeysuckle	*Diervilla lonicera*		
Horsechestnut	*Aesculus hippocastanum*		
Inkberry	*Ilex glabra*		
Kentucky coffeetree	*Gymnocladus dioica*		

Table 6 - 5

SPECIES THAT ARE FAVORED FOOD FOR LATER GYPSY MOTH
LARVAL INSTARS

GROUP 2

Common Name	Scientific Name
Beech, American	*Fagus grandifolia*
Cedar, red	*Juniperus virginiana*
Hemlock, eastern	*Tsuga canadensis*
Pine	*Pinus*
jack	*P. banksiana*
pitch	*P. rigida*
red	*P. resinosa*
Scotch	*P. sylvestris*
white, eastern	*P. strobus*
white, western	*P. monticola*
Virginia	*P. virginiana*
Spruce	*Picea*
black	*P. glabra*
Norway	*P. abris*
red	*P. rubens*
white	*P. glauca*
Plum	*Prunus*
Allegheny	*P. alleghaniensis*
American	*P. americana*
Canadian	*P. nigra*

Table 6 - 6

SPECIES THAT ARE MODERATELY FAVORED FOOD BY
LATER GYPSY MOTH LARVAL INSTARS

GROUP 3

Common Name	Scientific Name
Bayberry	*Myrica*
common	*M. carolinensis*
European, sweet gale	*M. gale*

Birch	*Betulla*
black	*B. lenta*
yellow	*B. alleghaniensis*
Blueberry	*Vaccinium*
low	*V. pennsylvanicum*
tall	*V. corymbosum*
Butternut	*Juglans cinerea*
Cedar, southern white	*Thuja occidentalis*
Cherry	*Prunus*
black	*P. serotina*
chokecherry	*P. virginiana*
Mazzard (sweet)	*P. avium*
pin	*P. pennsylvanica*
Elm	*Ulmus*
American	*U. americana*
English	*U. procera*
rock	*U. thomasii*
slippery	*U. rubra*
Hickory	*Carya*
bitternut	*C. cordiformis*
mockernut	*C. tomentosa*
pignut	*C. glabra*
shagbark	*C. ovata*
Hackberry	*Celtis occidentalis*
Hornbeam, American	*Carpinus caroliniana*
Hophornbeam, eastern	*Ostrya virginiana*
Maple	*Acer*
Norway	*A. platanoides*
red	*A. rubrum*
silver	*A. saccharinum*
sugar	*A. saccharum*
Pear, common	*Pyrus communis*
Poplar, silver	*Populus alba*
Sassafras	*Sassafras albidum*
Sweetfern	*Myrica asplenifolia*
Sweetgum	*Liquidambar styraciflua*

Table 6 - 8

OAK RIDGE DEFOLIATION IN MONROE COUNTY, PENNSYLVANIA,
1969–1971

Tree Species	Number of Trees per Acre	Number of Trees per Acre		
		Alive	Declining	Dead
Red Maple	164	155	9	0
Black Gum	135	135	0	0
Sweet Birch	13	13	0	0
Tuliptree	3	3	0	0
Chestnut Oak	70	0	28	42
White Oak	4	0	2	2
Red Oak	1	0	1	0

(Courtesy: Bitzer, 1971)

Table 6 - 9

TREE MORTALITY DAMAGE APPRAISAL IN PIKE AND MONROE
COUNTIES, PENNSYLVANIA, 1972

Tree Species	Sawtimber, 1,000 Board ft.°			Pulpwood, 1,000 Cubic ft.°		
	Healthy	Declining	Dead	Healthy	Declining	Dead
White Oak	7,524	2,249	1,104	1,564	1,948	994
Red Oak	931	1,961	686	417	530	221
Chestnut Oak	767	1,412	1,021	1,266	1,823	2,420
Red Maple	413	102	—	2,377	437	73
Black Birch	150	25	—	319	66	70
Hemlock	—	—	—	—	—	10
Sassafras	—	—	—	274	60	16
Pitch Pine	1,294	—	—	319	26	39
Miscellaneous°°	335	14	—	1,105	55	10

° trees twelve inches or more in diameter
°°black gum, aspen, beech, hickery, white pine

(Courtesy: Quimby, 1972)

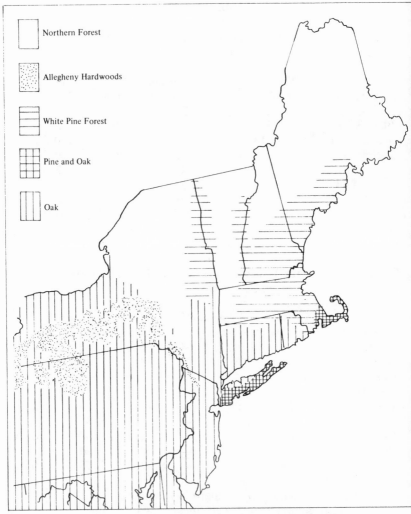

Major Forest Regions of Northeastern United States. *Courtesy:*
Massachusetts Forest and Park Association, after Behre, et al., 1936
Gypsy Moth Defoliated Areas, 1973. *Courtesy: USDA, 1973b*

Legend:
- Northern Forest
- Allegheny Hardwoods
- White Pine Forest
- Pine and Oak
- Oak

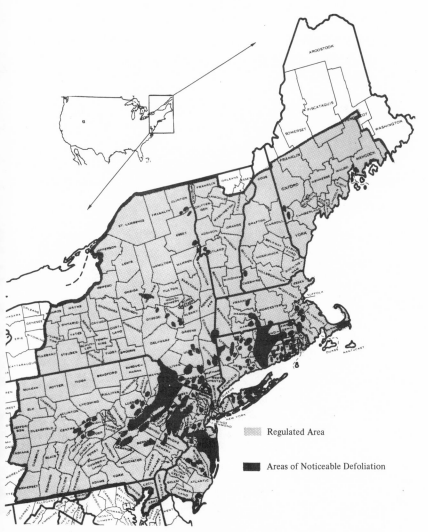

GYPSY MOTH DEFOLIATED AREAS, 1973. *Courtesy: USDA, 1973b*

7

Physiological Stress to Trees as a Result of Defoliation

The impact of gypsy moth defoliation on woodlands is illustrated in the nature, magnitude, and significance of the physiological response of the trees to defoliation. Stress on individual trees depends upon the species attacked, season of the year, climatic conditions, duration and degree of defoliation, stand site, location of defoliation in the tree crown, and the age and health of the tree at the time defoliation occurs. The most serious effect of gypsy moth defoliation is a series of "growth impact" damages including mortality, growth loss, and increased susceptibility to secondary insects and diseases. Defoliation also affects seed and flower production, epicormic branching, wood quality, foliation timing, respiration and efficiency of photosynthesis, size, abundance and food storage capacity, metabolic food conversion pathways, and hormone production. These components of forest productivity and forest tree physiology are currently being studied by the United States Forest Service, which requires the monitoring of defoliation over long periods of time.

The exact procedures causing mortality and physical injuries from damaged internal processes are not completely understood. With respect to the time of defoliation, it has been demonstrated that removal of new foliage decreases growth of the upper stem more than in the lower stem, whereas the reverse is true when old foliage is removed (Craighead, 1940). In addition, late season defoliation reduces twigs of woody tissue and subjects them to winter damage (Kramer and Wetmore, 1943).

Tree mortality is often the result of secondary agents, such as bark beetles, borers, and fungi. The most important of these secondary agents are the root rot fungus or shoe-string root rot fungus, *Armillaria mellia*, and the two-lined chestnut borer, *Agrilus bilineatus*. The chestnut borer invades weakened trees and constructs tunnels under the bark that stop the flow of water and nutrients in the tree. Wargo (Marx, 1970b) has indicated that the action of *A. mellia* and *A. bilineatus* in the outer wood may be controlled by the chemical changes occurring there as a result of defoliation. Many secondary insects propagate in dead and weakened trees, then attack and kill trees that otherwise would have survived.

Mortality often is affected by site. Repeated defoliation increases mixed hardwood mortality on dry sites (Stephens, 1971), while mixed hardwood stands on wet sites are more resistant to gypsy moth attack.

Specific growth losses from defoliation include reduced height growth, radial growth losses, and reduction of shoot elongation. Defoliation of pines often results in reduction in the dry weight of shoots and reduced length of new needles. Incomplete and missing growth rings also may accompany defoliation (Kulman, 1965; and O'Neil, 1963).

The loss of foliage probably has a proportional effect on hormone production, stored products, photosynthate production, and an indirect effect on osmotic pressure. Defoliation also affects the translocation and distribution of photosynthates and hormones (Kozlowski, 1969; and Kulman, 1971).

Stomatal resistance to transpiration is also affected by defoliation. This resistance is less in fully emerged, secondary foliage produced after defoliation than in primary foliage that has escaped defoliation in oak and aspen; however, accompanying the loss of leaf surface and hence the area for evaporation, the demand for moisture decreases (Stephens et al., 1972). The decreased demand for moisture improves the internal water balance of defoliated trees if ample soil water is available (Stephens et al., 1972); but defoliation also may increase tree hydration, as is the case with the red pine (Waggoner and Turner, 1971).

Defoliation of evergreens may cause a loss of stored food, storage places for food, and may affect the production of growth regulators. Drought compounds the effects on the physiological make-up of trees during and after defoliation. Stress factors in periods of drought effect greater reduction in root starch and reducing sugars, and some acids are higher in concentration in parched plants than in non-defoliated plants (Houston, 1973).

8

Control of the Gypsy Moth: Introduction

There are several natural factors that aid in controlling gypsy moth populations. Winter temperatures below $-29C°$ kill unprotected eggs; late spring frosts kill newly hatched larvae; parasites, predators, and diseases attack all four life stages; and starvation helps in reducing overpopulated communities of the gypsy moth.

Although these natural control factors lengthen the time intervals between gypsy moth impact population peaks, they do not significantly prevent the periodic and severe outbreaks of such impact populations. Therefore, in order to prevent such populations and to reduce their damage, and control the gypsy moth to a significant level, applied forest management control measures are used against the insect. Control refers to any applied method or combination of methods that effects the reduction or prevention of damage by the insect. Applied measures refer to such direct methods of control as biological, chemical, genetic, physical, radiation, and regulatory, and to the silvicultural technique, an indirect control measure.

Regardless of the method or methods chosen, the applied control must kill a significantly large percentage of the pest population in order to be effective.

9

Chemical Control

At the present time the main weapons against the gypsy moth are a variety of nonpersistent, synthetic chemical insecticides. However, the contamination of the world's water and land resources with persistent chemical pesticides has resulted in an ever-increasing concern about the dangers of chemical pesticides in general. Since the publication of Rachel Carson's *Silent Spring* in 1962, the general public has become highly aware of the presence of chemical insecticides and their harm to the biosphere. In fact, public interest in the problems and dangers of chemical insecticides made such a giant leap after the publication of *Silent Spring*, that it culminated ten years later in a near-total ban of DDT, once hailed as the farmer's hero. In 1974 the Environmental Protection Agency banned the use and manufacture of aldrin and dieldrin, two widely used pesticides in the same chemical family as DDT. The production, sales, and use of two more pesticides, chlordane and heptachlor, used in household products and on various agricultural products, were banned by the EPA in 1975. The reason for this extensive ban is the extremely high cancer risks these pesticides present. Today the EPA is considering the prohibition of 2, 4, 5-T and endrin. The reaction to pesticides on the state level is also dramatic. Connecticut, for example, took further steps in 1972 against the use of pesticides by banning their application by air (Marx, 1972b).

Insecticides of the types now available, persistent and nonpersistent, present a number of complex hazards. These hazards, more closely associated with the persistent types of chemical insecticide, involve chemical residues in foods and wildlife resources, adverse effects upon insect pollinators, parasites, and predators, food-chain magnification of insecticides, and the ability of pest insects to become resistant to the insecticides. Certain species of insects have become resistant to all of the following insecticides: DDT, cyclodienes, OP (organic phosphate) compounds, and carbamates. These species include the house fly, *Musca domestica,* the sheep blow fly, *Melophagus ovinus,* and the cotton leafworm, *Alabama argillacea.* In addition, by 1967, 224 species of insects and acarines, 127 of which attack field or stored products, had developed resistance to pesticides (Brown, 1967). Many more insects are still in stages of insecticidal-resistance development.

The basic principle of chemical or insecticidal control of the gypsy moth is to apply an insecticide to the young foliage in sufficient quantity to kill the feeding caterpillars. The insecticide is applied approximately one week after the eggs begin to hatch. Spraying by plane or helicopter is used to treat large areas of woodlands. Application from the ground by mist blowers or hydraulic sprayers is used for controlling the insect on shade trees and around houses built in woodlands. Most spraying is done to prevent defoliation of trees heavily infested with the insect.

Knipling (1968) has suggested several general roles of conventional chemical insecticides applicable to the gypsy moth, as follows: (1) the control of an insect pest on a temporary basis; (2) the destruction of some insect pests immediately and effectively; (3) the ability of individual property owners to take action against an insect pest; (4) a favorable cost-benefit ratio provided by chemical insecticides to achieve control; and (5) the ability to select insecticides on the basis of their toxic action.

Whether pest control be on a large or small scale, the insecticide used must be considered before the program is initiated. The following factors, as suggested by the National Academy of Science (1971), should be reviewed: the effects of the insecticide against the target insect; possible hazards to applicators and residents in the treated area; food residues; hazards to wildlife, pollinators, and beneficial insect parasites and predators; effects of the pesticide on plants and soil organisms; possible water pollution and contamination of surrounding areas through air or water movements; absorption and translocation of pesticides by plants; relative persistence of the pesticide in the soil and water; and the action of the pesticide on the soil. From these factors it is understandable why the use of chemical pesticides is often hazardous and complex.

In 1897 lead arsenate, $PbHAsO_4$ (diplumbiohydrogen arsenate) was first used as a chemical pesticide against the gypsy moth in Massachusetts, and by 1938 it was the most effective spray material in use against the insect. Lead arsenate is an inorganic insecticide that acts as a stomach poison and is toxic to animal life. However, it is not toxic to gypsy moth parasites or predators, but it does kill bees that drink contaminated water or feed on contaminated pollen (Eckert, 1949). It is highly toxic to gypsy moth caterpillars that eat sprayed foliage. Arsenic-containing compounds combined with thiol (–SH) groups required by some enzymes thus interfere with enzyme action and cause cessation of their function.

Lead arsenate is usually applied with fish oil or raw linseed oil, which serve as adhesives to make the spray adhere firmly to the foliage. Sometimes skim milk is used as a spreader of sticker. For each pound of lead arsenate used, 20 gallons of water and four fluid ounces of oil are mixed into a solution. The mixture must be continuously agitated, even while the pesticide is being aplied. Once the compound is applied, it is unsafe to allow animals to feed upon the sprayed foliage.

In 1949 lead arsenate was replaced by synthetic organic pesticides due to the insecticide's high mammalian toxicity, its increasing problem of insect resistance, and its low efficiency of kill compared to the new synthetic, organic pesticides.

DDT (1), 1,1,1-trichloro-2,2-bis(p-chlorophenyl) ethane, first appeared in the United States in 1942 and was used in aerial-application programs against the gypsy moth in 1944. It was used *exclusively* from 1946 until 1963. DDT appeared

to be an extremely efficient weapon against the insect when applied at a rate of one pound per acre. Poisoning in insects by DDT and other chlorinated hydrocarbons is associated with disturbances in the central nervous system that result in hyperactivity, tremors, and uncoordination. The exact mechanism of this action is not known.

A major advantage of DDT is its ability to remain on foliage long enough to kill larvae that hatch over extended periods of time. In addition to being a persistent pesticide, the costs of DDT spraying programs are relatively inexpensive. In 1958 concern over possible environmental contamination caused a gradual transition from DDT to less persistent insecticides. DDT was eventually discontinued as a gypsy moth control measure by many state and federal agencies after 1963 because of public pressure, increasing treatment costs, and lack of funds. Pressure against the use of DDT continued as scientists began to find disturbing evidence concerning its environmental effects. Moreover, the gypsy moth was developing a tolerance for relatively large doses of DDT (Forgash, 1973).

In June of 1972, after 17 months of investigation, EPA Administrator William Ruckelshaus imposed an almost total ban on DDT. This ban has been challenged repeatedly by manufacturers and processors of DDT, and in 1974 the USFS requested permission to use DDT for a single, concentrated attack on the tussock moth in northwestern United States. All such challenges and requests have been denied. Organized letter campaigns requesting the spraying and reinstitution of DDT have been directed at many state and federal agencies. Even with the possible legislation being considered in Congress allowing restricted usage of DDT, many state agencies refuse to use the insecticide to combat the gypsy moth. The Pennsylvania Department of Environmental Resources has cited several reasons for their position against the renewed use of DDT (Hope, 1973), including the following: (1) Since the entire state is considered infested with the insect, several million acres would have to be treated with DDT to assure only two or three years of slight gypsy moth defoliation; (2) if the insect were controlled in this state, there would be the possibility of immediate reinfestation from neighboring infested states; and (3) gypsy moth parasites and predators would also be destroyed with the use of DDT (Hope, 1973). These parasites could not continue on their own natural control level if DDT were applied in a large-scale program.

On large, treated areas DDT is most effective when mixed in an oil solution at a concentration of 12% and applied at a rate of one pound of DDT per gallon per acre. To achieve this oil solution, DDT must be dissolved in an agent such as Sevacide 544 B and then diluted to a 12% solution. On small areas a 50% wettable powder of DDT is applied at a rate of two pounds of DDT to 100 gallons of water. In hand sprayers two tablespoons of DDT are added to each gallon of water.

The effectiveness of DDT against the gypsy moth can be measured by the many programs conducted by Pennsylvania from 1944 to 1963. During this 20-year period, DDT was used to treat nearly 2,000,000 acres of woodands, and its use helped to confine the insect almost entirely to the six northeastern counties of the state. Today, 14 years after the discontinuance of DDT by Pennsylvania, all 67 counties of the state are infested with the gypsy moth.

From the introduction of the gypsy moth into North America until the mid 1950s, many individuals and agencies believed that the insect could be eradicated by chemical means, especially with DDT. In 1957 approximately five million dollars was appropriated by state and federal governments for gypsy moth control work. It was during this year that the largest, single aerial spraying program ever conducted in the United States by the USDA was launched in an attempt to eradicate the gypsy moth from almost three million acres of woodlands in parts of New Jersey, New York, and Pennsylvania (USDA, 1957). DDT was applied at a rate of one pound of insecticide per gallon of light oil per acre. This eradication program failed.

Again, in 1958, Pennsylvania attempted to eradicate the gypsy moth with aerial application of DDT. This eradication program also failed. By 1958, after 15 years of DDT usage, approximately 680,000 acres had been treated in Pennsylvania; and by 1959 after an enormous eradication attempt in the state, over 1,000,000 acres had been treated with DDT at a time when gypsy moth eradication was believed possible. However, after the failure of 1959 in Pennsylvania and others before it, the insect's eradication was considered impossible.

Over recent years the gypsy moth has become highly

resistant to DDT; further, DDT penetrates the cuticle of the common house fly, *Musca domestica,* faster than it penetrates the cuticle of the gypsy moth (Tomlin and Forgash, 1972a and 1972b). For the most part, DDT was abandoned in 1964.

Protection from the gypsy moth as provided by DDT can easily be seen in this photograph. Trees on the right side of the road were not sprayed with DDT and have been defoliated by the insect, while those on the left were treated with DDT and show no signs of defoliation. *Courtesy: USDA*

Spraying operations conducted against the gypsy moth from 1964 to 1976 have been largely with nonpersistent insecticides. Among the many pesticides available and legally labeled for use against the insect are carbaryl (Union Carbide's Sevin) and tricholorfon (Chemagro's Dylox).

Carbaryl and trichlorofon are both capable of killing 90% or more of gypsy moth larvae in an infestation. Because they are nonpersistent, two or three sprayings are applied to obtain results comparable to DDT. Since eggs may hatch over a one-month period, increased spraying to cover this time period has resulted in higher costs and increased public criticism. Nevertheless, these insecticides are available commercially for gypsy moth control.

Carbaryl (2), 1-napthyl-N-methycarbamate, is a broad-spectrum, synthetic, organic-contact insecticide belonging to the carbamate chemical family. Introduced in 1958 it was the first commercial carbamate insecticide to be used on a large scale in the U.S. As a carbamate ester, its mode of action is analogous to that of organiphosphates, that is, it

Carbaryl (2)

$$0 - C - NH - CH_3$$
$$\overset{\|}{0}$$

acts as an acetylcholinesterases inhibitor. Following carbaryl ataxia and the appearance of paralysis in the larva's voluntary muscles, the insecticide attacks the central nervous system, leading to asphyxia and cardiac failure.

Carbaryl has relatively low toxicity in mammals, birds, and aquatic life; however, it is fairly toxic to bees, parasitic wasps, and beetle predators. Carbaryl has been shown to kill honey bees, *Apis mellifera* (Morse, 1961; and Strange et al., 1968), the parasitic wasp, *Brachymeria intermedia* (Tomlin and Forgash, 1972a), and the predaceous beetles, *Calosoma sycophanta* and *C. frigidum* (Doane and Schaefer, 1971). Carbaryl insecticide is listed along with 65 other pesticides as being highly toxic to bees (Atkins et al., 1970). Because carbaryl also kills beneficial insects, there is widespread opposition to its use. Experimentation with bee pollen traps, however, has offered some hope of reducing bee losses in areas treated with carbaryl (Marx, 1974b).

Trichlorofon (3), dimethyl (2,2,2-trichloro-1-hydroxyethyl) phosphate, is an aliphatic derivative of a phosphorus compound and is registered by the EPA for use against the gypsy moth. Its mode of operation is as an acetylcho-

$$(CH_3)_2 - \overset{\overset{\text{O}}{\|}}{P} - O - \overset{\overset{\text{OH}}{|}}{\underset{\underset{\text{H}}{|}}{C}} - \overset{\overset{\text{Cl}}{\diagup}}{\underset{\underset{\text{Cl}}{\diagdown}}{C}} - Cl \qquad \text{Trichlorofon (3)}$$

linesterase inhibitor; recommended dosage is one pound of active ingredient per acre. Evaluation of trichlorofon as an insecticide for use in pest control programs began in 1961, and in 1973 the insecticide was used by state and federal agencies against the gypsy moth. The Pennsylvania Department of Agriculture and the EPA utilized trichlorofon in their 1973 spraying programs, and again in 1974, 1975, and 1976 because of the insecticide's safety with respect to pollinating insects. Trichlorofon is relatively nontoxic to forest and cropland environments, wildlife, beneficial insect parasites, predators, and pollinators.

Field tests conducted in 1971 and 1972 against the gypsy moth with aerially applied trichlorofon in oil carriers revealed no discernible effects on fish or other aquatic life. Further, the insecticide apparently underwent complete biological breakdown in water after five days (Grimble, 1973). Aside from very minimal contamination, trichlorofon in oil is one of the safest insecticides developed for aerial spraying at a dosage rate of one pound per acre.

Because the insecticide does not threaten bee populations, many agencies use it in place of carbaryl. In test areas sprayed with trichlorofon, no honey bee damage has been reported, and damage to acquatic insects is minimal (Marx, 1973b). Included in this test area were zones containing dozens of colonies that were sprayed twice with trichlorofon.

Because of nonpersistent nature of the chemical insecticides currently used against the gypsy moth, these pesticides are considered as only temporary stop-gap measures to relieve serious public-nuisance problems, having their most effective results on relatively small acreages containing high-value trees in high-public-use areas. There are several reasons to support the use of these insecticides as only temporary stop-gap measures: (1) Since the pesticides do not effect complete kill, the surviving insects repopulate the treated area in two to three years; (2) the dispersal of caterpillars from untreated to treated areas will continue gypsy moth infestation; (3) dispersal of larvae from the treated area before control is implemented serves as a means of reinfestation by a dispersed, reservoir population, and (4) areas missed during spraying operations also serve as reservoir populations from which reinfestation can occur.

The problems of chemical usage are further complicated and hindered by increasing treatment costs, lack of funds, public criticism, and a general anti-pesticide attitude on the part of the public.

Spraying large areas requires proper timing and careful preparation if adequate coverage is to be obtained. The insecticide must be applied when the insect is most vulnerable (during larval development) and when the insecticide will offer maximum protection to the trees. Massive spraying programs require aerial application of insecticides; coverage by proper aerial applications is aided by overlapping effective swaths or passes. Efficiency is largely dependent on droplet size of the pesticide, the optimum being the smallest size droplet that will reach the trees before evaporation (Balch, 1958).

There are several reasons why aerial applications may fail. The insecticide may be improperly prepared; spraying operations may be performed unsatisfactorily, having open or missed swaths; or the spraying may have resulted in insufficient and highly variable spray deposits. The treated area may have experienced unusually heavy precipitation after treatment; the insecticide and/or sticker may have failed to adhere to the foliage; or spray residues may dissipate from preferred host foliage.

Aerial spraying with current insecticides such as carbaryl and trichlorofon offers a practical means for treating large parks, developed woodlands with dwellings, and large camping sites. Even though many people still advocate widespread woodland spraying to solve the gypsy moth problem, however, massive aerial spraying is not supported by state and federal agencies. The Pennsylvania Department of Environmental Resources, for example, opposes the use of massive chemical-spraying progress to combat the gypsy moth.

Accompanying the change from persistent chemical insecticides such as DDT to nonpersistent sprays is a change in the concept of spraying programs against the gypsy moth —from eradication and containment attempts to temporary stop-gap measures providing protection of high-value trees and selected areas until other agents are perfected for control.

Many areas in the northeastern United States sprayed from 1969 to 1976 have been heavily defoliated due to rapid population increases in surviving insects. This has occurred for several reasons: The insecticides could not cope with larvae that hatched over a period of one month unless two or more applications were made; the amount of time required for effective treatments and the size of many of the infested areas hindered the logistics of effective, large-scale spraying oprations; areas missed during spraying operations due to faulty overlapping of spray swaths and omission of trees around lake shores and streams served as reservoirs for population outbreaks of the insect; and many larvae in dense populations dispersed from these infestations before spraying occurred. These larvae helped to ensure continued infestations. Because of migrating and wind-borne dispersing larvae, it was also difficult to construct treatment boundaries.

From 1964 to 1971 the Pennsylvania Department of Agriculture treated all known infestations in the state with registered insecticides. However, this did not prevent the gypsy moth from spreading across the state. This and numerous other examples demonstrate that although current insecticides may yield 90% kill, their nonpersistent nature and the population dynamics of the gypsy moth render them relatively ineffective in achieving significant control of the insect (Pennsylvania Department of Environmental Resources, 1973).

Currently, there is no known insecticide to solve the gypsy moth problem or to prevent its further spread to the south and southwest. Present chemical insecticides being considered for use against the gypsy moth are screened by the USDA, Animal and Plant Health Inspection Service at Otis Air Force Base, Massachusetts. Under simulated field conditions, a series of tests is conducted on each insecticide to determine its toxicity to the gyspy moth. These tests are also used to find safer and more effective insecticides.

There are three basic tests conducted by the USDA; these include the artificial diet, and epidermal and seedling tests (McLane, 1973). In the artificial-diet test method, the insecticide is mixed with an artificial food and exposed to the second-instar larvae. Twenty-four-hour mortality readings are taken for a three-day period against a control group, and the insecticide is screened as a stomach or contact poison. In the epidermal test method, an insecticide/acetone mixture is applied directly to the larvae with the aid of a syringe. A mortality reading is made 48 hours later and compared with a control. In the seedling method, young plants are treated with an insecticide and then subjected to larval feeding; mortality readings are taken every 24 hours for a three- to six-day period.

During 1972 the USDA screened approximately 75 toxicants and is presently in contact with 30 companies in an attempt to establish rate, dosage, aging, and weathering information on some of the more promising and newer chemical compounds (McLane, 1973).

Independent laboratory screening tests have revealed many chemical insecticides equal to or better than carbaryl against young gypsy moth larvae. The following insecticides are included in that category: 0,0-diethyl 0-3,5,6-trichloro-2-pyridyl phosphorothioate, diazionon, dimethrin, 3,4,5-trimethylphenyl methylcarbamate, and benzo(b)thien-4-yl methylcarbamate (Merriam et al., 1970).

Furthermore, tests with natural pyrethrins demonstrate that there is great potential in using them to control the gypsy moth (Dunbar and Doane, 1973). Pyrethrum (4) is a mixture of four pyrethrins principally obtained from the plant *Chrysanthemum cinariaefoliuna.* Although pyrethrum

Pyrethrin (4)

is an expensive insecticide and susceptible to photo-decomposition, it is nonpersistent and possesses low oral mammalian toxicity and rapid insecticidal action. New pyrethroids are being developed that are stable for four or five days (Sanders, 1975). The exact method of pyrethrum action has not yet been elucidated.

During the last three decades an increased heavy reliance upon the use of chemicals for suppression of most insect pests has occurred. However, chemical insecticides have their limits, and many would prefer to use less drastic measures such as biological control, pheromone control, or sterilization and genetic control programs to achieve partial or complete protection and relief from the gypsy moth.

TWELVE REASONS AGAINST THE USE OF MASSIVE CHEMICAL SPRAYING PROGRAMS TO COMBAT THE GYPSY MOTH

1. There is no known registered insecticide that will solve the gypsy moth problem or prevent its further spread.
2. Natural windspread of early instars often occurs before there is sufficient tree-leaf development to institute spraying operations.
3. Current nonpersistent insecticides registered for gypsy moth control provide approximately 90% control. Within two or three years following treatment, however, the remaining populations can repopulate the infested areas to their former level.
4. Those areas missed while spraying serve as reservoirs for new outbreaks. The resulting reinfestation would lead to a never-ending and escalating program of insecticide sprayings.
5. To interfere with the normal collapse of a gypsy moth population following severe tree defoliation is self-defeating.
6. Because the gypsy moth extends its hatching period up to two months, it is impossible to obtain better control using present insecticides that are effective only for one or two weeks.
7. Massive spraying is indiscriminate and results in blanket coverage of large areas. This often results in contamination of streams, ponds, and other "hazardous" areas.
8. Weather often disrupts spraying programs. Excessive rain or cold weather during the spraying season may reduce the efficiency of the operation.
9. Often the strongest larvae survive such spraying operations and these individuals may create a worse problem later.
10. Massive chemical spraying may cause considerable damage to gypsy moth parasites, predators, and many insect pollinators.
11. Massive spraying programs usually are unacceptable to the public; also, the costs of such programs are extremely high.
12. The overabundance of solid oak stands in many parts of the Northeast is a major reason why the gypsy moth is so serious a threat. If this forest condition is maintained by chemical sprayings, it is possible that the gypsy moth problem may never end.

(Courtesy: Pa. Dept. of Envir. Res. 1973)

Table 9 - 10

POTENTIAL GYPSY MOTH INSECTICIDES

INSECTICIDE	DOSAGE EQUIVALENT lbs/acre	PERCENT MORTALITY AFTER INDICATED DAYS			
		1	2	3	4
Second-instar larvae					
Carbaryl	0.500	60	88	94	
Dursban	.500	86	100		
	.125	82	100		
Diazinon	.500	88	96	100	
	.125	66	88		
SD-8530	1.250	50	86	100	
	.125	34	96	100	
SD-9098	.500	78	98		
	.125	76	98	100	
MC-A-600	.625	68	86	100	
Dimethrin	3.300	52	92	98	
Tetramethrin	.200	88	92	94	
Fourth-instar larvae					
Carbaryl	0.250	50	54	74	86
Dursban	.040	52	60	72	74
Trichlorfon	.125	54	88	98	100
Diazinon	.125	34	56	64	70

Mortality of second-instar gypsy moths in laboratory insecticide screening tests, 1966–1967, and of fourth instars in 1967. Chemical insecticides shown equal to or better than carbaryl against young gypsy moth larvae in these tests included dursban, diazinon, dimethrin, Shell SD-8530, and Mibol MC-A-600.

(Courtesy: Entomological Society of America, from Merriam et al., 1970)

10

Sterilization Control

If costs can be reduced and techniques improved sterilization programs against the gypsy moth may provide significant control of the insects.

There are two basic sterilization methods used to achieve control or suppression of a pest population. The first technique involves the collecting and mass rearing of insects that are sterilized and then released into the environment. Methods of application used to cause sterility include exposure to atomic radiation, exposure to chemosterilants, crossing related strains to produce hybrid sterility, and releasing related strains that carry lethal traits. The second method consists of treating a portion of the natural pest population with a chemosterilant.

With respect to gypsy moth control by sterilization, only three techniques warrant discussion as possible control measures: sterilizing collected insects, crossing related strains to produce hybrid sterility, and treating the pest population with a chemosterilant.

Irradiation as a practical means of pest control was established in 1955 when the screwworm fly, *Callitroga hominivorax*, was eradicated from the West Indian island of Curaçao by releasing 400 irradiated males each week for 22 weeks for each square mile of the island (O'Brien and

Wolfe, 1964). Since then, exploratory studies with radiation control against other insects have occasionally been encouraging. The sterility technique of radiation control is not practical for controlling some insect pests, but it is hoped that it may play a prominent role in the control and regulation of the gypsy moth.

The sterility principle of control can be applied through the sterile-male or rear-and-release technique. The basic concept of this control procedure is that the sexually sterilized insects are allowed to mate with normal insects in the natural population, thus neutralizing the reproductive potential of the natural population. If the number of sterile or "neutralized" matings significantly exceeds the number of normal matings in each generation, the natural population will decreased in each subsequent generation. This is accomplished by maintaining a constant number of sterile insects in the natural population through seasonal or yearly releases of sterile insects, while the number of fertile insects declines. Thus the ratio of sterile to normal matings increases in each successive generation, resulting in the reduction in numbers of the natural population and perhaps their eventual elimination.

Progress has been made with gypsy moth sterilization control, but there are problems. Poor sexual competition of irradiated males due to radiation damage and inability to produce sufficient male numbers when required are the primary troubles with this approach.

In most species, released sterile males play a more significant role than released sterile females. Such is the case with the gypsy moth. According to the type of reproductive cells present in the male at the time of sterilization, at least three different forms of sterility may occur. These include dominant lethal mutations in the sperm, aspermia, and sperm inactivation.

Sperm having radiation induced, dominant lethal mutations can fertilize an egg, but the resulting embryo cannot develop. Radiation treatment may render the male incapable of producing sperm cells—aspermia. This condition is due to the interruption of spermatogenesis by irradiation. The production of inactive sperm or nonmotile sperm also may result from radiation treatment.

In order to control the gypsy moth by the sterile-male technique, it must be possible to cultivate the insect on a large enough scale to outnumber its indigenous population in the particular infested area to be controlled. This may involve the rearing of millions of insects that are of known quality and free of contaminants. The gypsy moth is not mass reared easily for several reasons, the most important of which are the present dependency on field-collected eggs, which often are disease carriers, and the high mortality rate experienced during rearing programs. In spite of these problems, rearing techniques are improving. Over 75,000 male pupae were reared, irradiated, and released over seven acres of infested woodlands in 1972, yielding a sterile to fertile ratio of 86:1 (Stevens, 1973b). During rearing programs it is necessary to establish diel cycle compatibility with the existing conditions at the intended treatment area to produce sterile adults compatible with fertile adults II, page 132.

There also must be a procedure to induce sterility in the male with 100% certainty; yet, this procedure must not endanger thes longevity, vigor, and behavior of the male, or its ability to compete sexually with indigenous males. Since it is impractical to separate males and females before irradiation, procedure for radiation treatment also must ren-irradiated females incapable of producing fertile eggs.

Although radiation treatment is expensive, the sterile-male technique should be made as economically feasible and operable as possible on a large scale.

There should be an adequate release program for distributing the sterile males within the native population. To achieve this, it is necessary to know how many individuals are required per unit area and per unit time for any given infestation. It is also essential that the released insects do not in themselves exhibit a source of injury. Released male moths have been observed to travel in all directions for distances up to ¾ mile (Stevens, 1973b).

Through the programming of required light and temperature cycles, mass rearing of gypsy moth larvae can be accomplished on an artificial diet (Leonard and Doane, 1966). However, the control of viral and bacterial pathogens in mass cultures must be achieved. The usual source of radiation for inducing sterility is a radioactive cobalt 60 source, Co^{60}, which emits gamma rays.

A preliminary study in 1958 by the Brookhaven National Laboratory revealed that 100% mortality of eggs occurred when males were exposed to 20,000 roentgens; males exposed to this radiation treatment were so exposed when they were six days or less old during their pupal stage of development. Later studies showed that 99% sterilization of males occurred when pupae were exposed to 22,500 R to 25,000 R and that lower doses of radiation usually resulted in complete sterilization of the female moth, but incomplete sterilization of the male moth. However, these high dosages of radiation resulted in a high percentage of adult males with crippling defects, preventing them from competing sexually with indigenous males under field conditions.

Radiologically treated moths demonstrate a high incidence of wing malformation when pupae are exposed to 27,500 R (Knipling, 1970). In order that somatic damage in adults be minimized, pupae older than six days have been singled out as the target stage for sterilization treatment. While sterility results from the mating of males treated with 20,000 R, the sexual competitiveness of the male moth in the field is limited. Laboratory-reared males appear more docile than wild males (O'Dell, 1973). Hatching of eggs from normal females mated with males that had received 20,000 R as six- to eight-day-old pupae was reduced by 50%, a level hardly sufficient for field application (Waters and Rule, 1960). Attempts to achieve sterilization control of the gypsy moth are being conducted with the use of increased radiation exposure, 22,500 R and higher, and with the use of older pupae, eight days or older, as the target stage for treatment.

At exposure to 20,000 R from a Co^{60} source, the radiation intensity causes changes in the size, structure, and arrangement of cells in the walls of the testes and follicles, in the vas deferens, and in fat body cells. Rule (1961) has cited five very evident changes that occur in the gonads as a result of irradiation: (1) degeneration of germinal cysts; (2) rupturing of sperm bundles in the follicles; (3) blockage of the funnel mouth of the vas deferens in the testes by the formation of interstitial tissue; (4) thickening of the vas deferens and the testes walls and septa; and (5) clumping of fat body cells. These damages to the testes represent some possible causes of sterilization in the male gypsy moth after radiation treatment. Cytological abnormalities in the reproductive cells of the male include mitotic inhibition, chromosome stickiness, and chromosome aberrations in the form of fragments, exchanges, inversions, and translocations. These forms of cytological abnormalities represent sterilization in the form of dominant lethal mutations carried in the sperm cells.

Gross morphological damage resulting to moths whose pupal stages where exposed to 20,000 R are crumpled-wing formations and hairless body development. To what degree genetic damage caused by irradiating pupae of the parental generation is transmitted from nonsterile, radiation-exposed males to the filial generation is unknown; however, it is possible that sterility or mutations may be transmitted to the offspring. Genetic damage caused by irradiating adults and pupae of the parental generation of the Indian meal moth, *Plodia interpunctella,* is sometimes transmitted to the first generation, where it produces an increased incidence of sterility and a reduction in total progeny (Cogburn et al., 1966).

Although the use of sterilization through radiation techniques is currently not a major component of federal or state programs combating the gypsy moth, it shows promise of larger involvement in population regulation and control of the pest. Extensive work in the development techniques of irradiation as applied to the gypsy moth are presently being conducted by the Plant Pest Control Division, Methods Improvement Laboratory, of the USDA at Otis Air Force Base.

The effectiveness of radiation control through the sterile-male technique can be seen in mathematical Models I and II, page 132.

The theoretical characteristic effect of the sterile-male technique on population trends in a population with a ten-fold increase per generation, such as the gypsy moth, is shown in Model I. With a starting population of one million fertile insects, the release of 49 million sterile insects would provide a sterile to fertile insect ratio of 49:1. Theoretically, this would nullify the reproductive capacity of 98% of the natural population. The remaining 2% that encounter fertile matings would produce a generation of 200,000, a 10-fold increase from the remaining 20,000 fertile insects. When the same number of sterile insects is released again in the second generation, the ratio of sterile to fertile insects would increase to 245:1, thus providing 99.6% control of the population. The resulting third generation would be 8,130 in number. With one more release of 49 million sterile insects into the third generation, the entire population would be eliminated due to the inability of fertile insects to find and mate with fertile insects.

As shown in Model II, the effectiveness of the sterile-male technique can be seen when compared to the control provided by chemical insecticides. When subjected to 98% kill with insecticides in each generation, it would take six years to achieve the same amount of control as the sterile-male technique for one million insects with a 10-fold increase potential per generation. This is based on insecticides that provide 98% kill; however, today's insecticides used against the gypsy moth provide roughly 90% kill. The reason for the longer period of time required by the insecticides to achieve control is their constant kill rate of 98% in this situation through each generation, while the effectiveness of the sterilization method increases with each generation, that is, 98% in the first generation, 99.6% in the second generation, and 100% in the third generation.

Furthermore, the negative aspects of chemical insectiides would be avoided with the use of the sterile-male technique. In contrast to insecticides, sterilization control leaves no residues in food or wildlife resources, does not affect insect pollinators, parasites and predators, and has no food chain magnification or other adverse effects upon the environment. However, sterilization control is much more expensive than chemical control and has not yet been perfected in gypsy moth control.

Many of the sterile-male release programs have not yet been able to maintain their control effectiveness or achieve their theoretical control. However, research work continues on sterilization techniques and release patterns for use in low-level population infestations of the gypsy moth. Field tests have indicated that a ratio of 40 : 1 in sterile-to-fertile males will suppress reproduction to a high degree on small plots (Knipling, 1970). Survival, dispersal, handling methods, and release programs also are continuously being evaluated and improved.

An alternative means of sterilization is provided through

the use of chemosterilants. Experimental work with the chemosterilants tepa, metap, and apholate was conducted in the early to mid-1960s by the USDA, Agriculture Research Service at Otis Air Force Base. Tepa (5), tris(1-aziridinyl)-phosphate oxide, a relatively inexpensive chemosterilant,

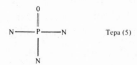

behaves as an alkylating agent and is radiomimetic, producing cytological damage including chromosome aberrations in the reproductive cells of the exposed individual. Collins and Downey (1967) have shown that the gypsy moth can be sterilized successfully with tepa without any adverse effects on mating competitiveness. It also was determined in the early 1960s that chemosterilized males were more competitive with wild males than irradiated males (Stevens, 1973a). An advantage of tepa is that it can be used in rear-and-release field programs. However, this chemical agent used experimentally between 1963 and 1968 was abandoned because of potential environmental problems (Stevens, 1973a), and Cobalt irradiation was substituted. Perhaps new formulations of tepa or designs of its possible usage may renew the use of the chemical agent or other similar chemosterilants in gypsy moth control programs.

Downes (1959) and Leonard (1973) have suggested the possible use of intermating strains of the gypsy moth to achieve genetic control of the insect. Goldschmidt (1934) found that gypsy moth females from the northeast part of the Japanese island of Honshu, when mated with males from western Europe or North America, produced females that were intersexed or sterile. The males resulting from this cross, when backcrossed with North American females, produced intersexed females one half of the time. Following the work of Goldschmidt, Downes (1959) proposed the release of Japanese gypsy moth males in infested areas of North America; however, the genetic basis of gypsy moth intersexes has not yet been fully established. Moreover, the resulting hybrid strain of Japanese-American mixture must be examined for its fecundity, range of host plants, and adjustment to climatic conditions in North America. If the resulting hybrid proves to be more vigorous than the present North American race, then the release of Japanese gypsy moth males would not be feasible. However, if this hybrid is acceptable, the use of the Japanese male insect for genetic control of the gypsy moth would have several advantages over the sterile male technique. This new method of control could be accomplished without irradiating the Japanese male insects, and there would be successful reductions in population numbers of the gypsy moth with greatly reduced operational costs.

Table 10 - 11

TREND OF AN INSECT POPULATION SUBJECTED IN EACH GENERATION
TO 98% KILL WITH INSECTICIDES

	NUMBER OF INSECTS IN POPULATION		
GENERATION	Before Treatment	After Treatment	NUMBER OF PROGENY
1	1,000,000	20,000	200,000
2	200,000	4,000	40,000
3	40,000	800	8,000
4	8,000	160	1,600
5	1,600	32	3200
6	320	3	320
7	30	1	0

(Courtesy: National Academy of Sciences, 1969)

Table 10 - 12

TREND OF AN INSECT POPULATION CONSISTING OF 1 MILLION
INDIVIDUALS SUBJECTED TO THE SUSTAINED RELEASE OF...
49 MILLION STERILE INSECTS

GENERATION	FERTILE POPULATION	REALEASED STERILE POPULATION	RATIO OF STERILE TO FERTILE	NUMBER OF PROGENY
1	1,000,000	49,000,000	49 : 1	200,000
2	200,000	49,000,000	245 : 1	8,130
3	8,130	49,000,000	6,027 : 1	0

(Courtesy: National Academy of Sciences, 1969)

11

Biological Control

When a pest insect is accidentally introduced into a new habitat from a foreign country, and natural enemies are absent from this new habitat, the resulting pest population may increase unchecked by natural controls. In Asia and Europe the gypsy moth is attacked by some 100 species of parasites and predatory insects. This in itself emphasizes the significance of importing natural enemies of the gypsy moth to North America for possible biological control. As the gypsy moth continues to spread from the northeastern United States, it has created an intense interest in various aspects of biological control.

Biological control is the regulation of plant or animal populations by natural agents. The action of parasites, predators, and pathogens helps in preventing populations of pest organisms from growing in an exponential manner. By doing so, these agents regulate to varying degrees the pest population's density at a lower level than would occur in their absence. These agents often benefit man indirectly by suppressing pest species below economically injurious levels.

Biological control is often referred to as any method involving the direct manipulation of natural enemies by introduction of or propagation by man. By 1958 the United States had imported 485 species of parasites and predators for use in pest-control programs against 91 insects, ten of which were forest pests (Balch, 1958). The most extensive effort to control an insect pest by biological means has been directed against the gypsy moth. This program began in 1906 and has been the most expensive ever attempted in the United States (Balch, 1958).

Introduced parasites and predators, if successful, gradually assume control season by season and generation by generation, with effectiveness varying in different climatic regions. If the parasite or predator has a narrower range of tolerance for climatic factors than its host, it will be effective only when these factors are favorable. The introduction of some parasites must be made annually in order to ensure continued effectiveness. Biological control aims at permanent results and tends to be self-perpetuating. Presently, researchers are exploring several ways to control the gypsy moth through natural means. They are attempting to produce fatal disease epidemics, to interrupt reproductive cycles, and to exploit predators and parasites.

As stated earlier, the gypsy moth is controlled in Europe and Asia by approximately 100 known insect parasites and insect predators. In the United States there also are several native predators and parasites known to attack the gypsy moth; however, these native agents do not appear to be effective in checking the insect's growth. Birds and small mammals feed on various stages of the gypsy moth, but they exert their greatest control only when gypsy moth populations are sparse. Moreover, the percentage of parasitism or predation is extremely variable within a defined infested area.

There are over 40 species of birds that feed upon the gypsy moth, but only a few are useful in destroying large numbers of the insect. Such birds are the blackbird, bluejay, catbird, black-capped chickadee, crow, black-billed cuckoee, yellow-billed cuckoee, grackle, nuthatch, oriole,

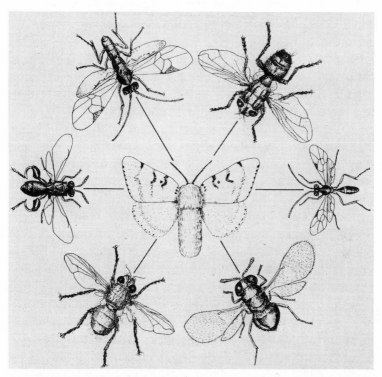

Shown are several gypsy moth parasites. From middle left and continuing clockwise are *Brachymeria intermedia, Apanteles melanoscelus, Compsilura concinnata, Phobocampe disparis, Eoencyrtus kuwanae,* and *Blepharipa scutellata.* In the center of the illustration is the gypsy moth female adult. *Courtesy: U.S. Forest Service*

136 HISTORY, BIOLOGY, DAMAGE, AND CONTROL OF THE GYPSY MOTH

robin, chipping sparrow, tananger, vireo, woodpecker, and red-winged blackbird. The black-billed and yellow-billed cuckoos have been found in large numbers in infested areas (Marx, 1973a). The impact of avian predators on gypsy moth populations is presently being studied through gizzard analyses.

Deer mice, *Peromyscus maniculatus*, white-footed mice, *P. leucopus*, short-tailed shrews, *Blarina brevicauda*, and gray squirrels, *Sciurus carolinensis*, eat gypsy moth larvae. A radio telemetry system has been designed to obtain quantitative mortality information on small mammalian predators of the gypsy moth. This radio system indicates the death of tagged animals and facilitates recovery of the individuals. When a tagged animal dies, his decreased body temperature triggers a temperature control transmitter implanted in the individual, which produces a radio signal that aids in the recovery of the body. This telemetry system aids in determining intrinsic factors, such as those influencing population dynamics of gypsy moth predators that can be corrected or altered during management practices to achieve effectiveness of the predators at predictable levels. Two designs from this program are the use of nesting boxes and supplemental food during cold-weather periods. Their usage helps to increase population sizes and survival rates of small-rodent populations, which may exert a stronger predatory influence on gypsy moth infestations.

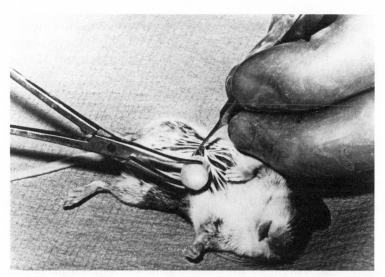

Movement patterns and mortality factors of the white-footed mouse, *Peromyscus luecopus,* an important predator of the gypsy moth, are monitored by a radio telemetry system. Part of this system consists of a battery-operated transistor signaling device, which is inserted into the mouse. *Courtesy: USDA*

Microbial diseases also attack gypsy moth populations. The two major microbes that attack the gypsy moth are the nuclear polyhedrosis virus, *Borrelinavirus reprimens,* and the spore-forming bacteria, *Bacillus thuringiensis.* (See also chap. 12, Microbial Control.) These two microbes play significant roles in causing an ultimate collapse in many gypsy moth infestations. In addition, *Streptococcus* bacteria often spread rapidly throughout outbreaks of the insect. Unfortunately, these later microbes have not yet been successfully adapted for large-scale management of the gypsy moth, although some progress is being made. *Streptococcus faecalis*, a small coccoidlike bacterium, has been used successfully by the Connecticut Agriculture Experiment Station, New Haven, in test spraying some foliage.

Eighty-six pathogenic, aerobic bacterial isolates have been obtained from diseased gypsy moth larvae in both sparse and dense populations (Podgwaite and Campbell, 1972). These bacterial isolates represent the families Achromobacteraceae, Bacillaceae, Enterobacteriaceae, Lactobacillaceae, and Pseudomonadaceae. Podgwaite and Campbell (1972) found *Streptococcus faecalis*, *Bacillus cereus*, *B. thuringiensis*, Group C *Enterobacter* types, and *Pseudomonas* spp. to be the most common pathogens. A number of additional pathogenic bacteria have been isolated from living and dead larvae, including *Proteus mixofaciens* and *Searratia marcesens* (Cosenza and Podgwaite, 1966; and Podgwaite and Campbell, 1972). An entomogenous fungus, *Paecilomyces farinosus*, also has been observed growing on egg clusters of the gypsy moth (Dunbar et al., 1972).

Podgwaite (1973) has suggested four basic factors contributing to the disease complex of gypsy moth larvae: (1) physiological-genetic stress (noninfectious) among early-instar larvae; (2) nuclear polyhedrosis incidence among early-instar larvae; (3) incidence of fungal disease in late-instar larvae; and (4) incidence of protozoan disease in all instars.

Protozoan and fungal diseases have not received so much attention as the bacterial or viral diseases of the gypsy moth. There are several microsporidian diseases of the gypsy moth larvae, but most of these protozoans are not native to North

America. In addition to *P. farinosus,* pathogenic fungi
isolated from gypsy moth larvae include *Beauveria bassiana,*
Metarrhizium anisopliae, and *Spicana forinosa* (Valisjevic,
1958).

There is a growing interest in viruses and, in particular,
cytoplasmic viruses for gypsy moth control. These viruses,
along with pathogenic bacteria, are being studied as mortality factors in the field and for their potential in biological
control programs. Many new viruses and bacteria are being
discovered and identified each year; some of them are
known to have a wide range of pathogenicity while others
are selective in nature, having strong specificities. The use
of these new pathogens in any control program depends
upon the gypsy moth's resistance to each, their dissemination
and propagation capacities, and their ability to cope with
climatic conditions. It is hoped that many may be used to
prevent or control population outbreaks of the insect.

Scavengers in the orders Collembola and Coleoptera and
in the families Sarcophagidae, Phoridae, Muscidae, and
Pentatomidae also have been found in dead larvae (Marx,
1973a).

Insect parasites and predators appear to be the most
important biological agents preventing and terminating outbreaks of the gypsy moth. There are several established
parasites abundant on eggs, larvae, and pupae of the insect.
Some native insects that have proved to be natural enemies
of the gypsy moth include ground beetles in the genus
Calosoma, such as *C. scrutator, C. calidium, C. wilcoxi,* and
C. frigidium. Also the fly parasite, *Exorista larvarum,* and
several species of the Pentatomidae are known to attack and
kill gypsy moth caterpillars. Ichneumonid wasps, *Theronis*
atalantae and *Itoplectis conquisitor,* have been found attacking the gypsy moth in the northeastern United States
(Smilowitz and Rhodes, 1972 and 1973). However, native
insect agents have been unable to control or check the gypsy
moth significantly in North America.

When the effectiveness of or the benefit derived from any parasite is evaluated, other factors including climatic conditions and forest stand and composition must be considered. The activities of parasites and predators are influenced by prey abundance (Solomon, 1949). Parasitoids and predators usually increase or decrease the percentage of their effect directly with host abundance, and both aggregate or disperse according to the population density of the host. These patterns of animal behavior are important to any biological control program since the program's success depends upon the density-dependent responses of the host's parasites or predators (Weseloh, 1973). Many factors also limit maximum parasitization. These are related to the parasite's dispersal and host-finding ability, microhabitat conditions, and population characteristics of the gypsy moth. The number of eggs, larvae, and pupae that a female parasite can reach with her ovipositor is also a critical factor (Weseloh, 1972). In some studies, significantly more female pupae than male pupae were parasitized (Tigner, 1973). In most studies the percentage of parasitization was generally highest when the gypsy moth population level was low (Campbell and Podgwaite, 1971; Weseloh, 1973).

Since the most effective natural enemies of the gypsy moth are lacking in North America, insect parasites and predators have for nearly 70 years been imported to the United States to combat it. In 1905 Massachusetts, together with the Federal Bureau of Entomology, began to import parasites and predaceous insects from Europe and Asia. Following the work of Howard and Fiske (1911), which stressed the importance of European and Asiatic insect parasites against the gypsy moth, importation programs were increased. By 1922 two predaceous ground beetles and nine species of parasites were established following the release of approximately 100,000,000 parasites into the American environment. Over 100 species of parasites and predators have been imported, reared, and released in infested areas of the United States for gypsy moth control. The more promising parasites are reared in large numbers and liberated in locations favorable for their development and populational growth.

Parasites and predators selected for use must have high reproductive potential, an ability to use alternate, non-beneficial hosts, a good searching or host-finding ability, and a life cycle synchronized with that of the gypsy moth. If parasites and predators are to be utilized as control agents, they must be proved effective under field conditions and must be nonpathogenic to man.

Today 13 major imported parasites are established in the Northeast. Of these, the most significant imported in the early 1900s are a predator beetle, *Calosoma sycophanta*; an egg parasite, *Ooencyrtus kuwanae* (wasp); five larval parasites, *Apanteles melanoscelus* and *A. porthetriae* (wasps), and *Compsilura concinnata, Parasetigena agilis*, and *Sturmia scutellata* (flies); and a pupal parasite, *Brachymeria intermedia* (wasp).

Calosoma sycophanta (Coleoptera : Carabidae) is a large, metallic beetle imported from Europe in 1906. This predator is more prey specific to the gypsy moth than its native relatives of North America in the genus *Calosoma*. Predation is conducted by the beetle in both its larval and adult forms, which feed on both caterpillars and pupae of the gypsy moth. Adult beetles may live for four or more years; each beetle is capable of consuming 100 to 150 gypsy moth larvae in a season, and the female is capable of laying 100 eggs.

Ooencyrtus kuwanae (Hymenoptera : Encyrtidae) is a multi-brooded, parasitic wasp that attacks the upper layers of gypsy moth eggs. A native of Japan, it was imported to North America in 1908 (Crossman, 1925). This tiny parasite is capable of producing five or six generations each year Each generation may attack the gypsy moth from late July until late December, when the wasp overwinters. The parasite is capable of parasitizing 40% of an egg mass.

Apanteles melanoscelus and *A. porthetriae* (Hymenoptera: Braconidae) were brought to the United States in 1911. Both are multi-brooded wasps that attack the early instars of the gypsy moth. The female in each species is capable of laying 1,000 eggs. There are usually two generations each year, and the wasps overwinter as full-grown larvae emerge from the host to form overwintering pupae. in transmitting the nuclear polyhedrosis virus.

Compsilura concinnata (Diptera: Tachinidae) is a parasitic fly that attacks gypsy moth larvae. Introduced into North America from Europe in 1908, the fly is larviparous, laying larvae inside the body wall of the gypsy moth caterpillar. The female is capable of laying 120 larvae in her lifetime and has over 200 alternate hosts throughout the northeastern United States. This fly also causes equally high parasitism in the satin moth, *Stilpnotia salicis*.

Parasetigena agilis (Diptera: Tachinidae) is also a parasitic fly native to Europe that attacks the larval stage of the gypsy moth. *P. agilis* was brought to the United States in 1908; the fly has one generation each year, but it is capable of laying over 200 eggs in its lifetime.

Sturmia scutellata (Diptera: Tachinidae) is a parasitic fly with one generation a year. However, the eggs of this fly are laid on the leaf of a host plant and consumed by late instars as they feed. The consumed eggs develop into larvae inside the instars and feed on the host caterpillar. The fly larvae emerger from the host to form overwintering pupae. The female fly is capable of laying 5,000 eggs in her lifetime.

Brachymeria intermedia (Hymenoptera: Chalcididae) is a parasitic wasp that attacks the gypsy moth pupae. The wasp is multi-brooded, having one complete generation and a partial second generation each year. The wasp is capable of parasitizing 75% or more of the pupae in heavy infestations.

Other ichneumonids, *Cocygomimus pedalis, Cratichneumon sp., Campoplex sp.,* and *Scambus sp.* have become established in some areas of the Northeast along with *Lespesia sp.* (Diptera: Tachnidae).

When in significant numbers, these major parasites stabilize gypsy moth populations following the insect's collapse. In New Jersey *P. agilis, S. scutellata, A. melanoscelus,* and *C. concinnata* have been found to be important in gypsy moth "stable areas" (Marx, 1972c).

In 1974 the Animal and Plant Health Inspection Service of the USDA released four new exotic species of gypsy moth parasites in the Northeast. These parasites included *A. liparidis* and *Meteorus pulchricornis* (Hymenoptera: Braconidae), imported from France and Morocco respectively, and *Paleyorista sp.* (Diptera: Tachinidae) and *Coccygomimus turionellae* (Hymenoptera: Ichneumonidae), both imported from India. *A. liparidis, M. pulchricornis,* and *Palexorista sp.* are larval parasites, while *C. turionellae* is a pupal parasite. Four additional larval parasites have been mass-reared by the New Jersey Department of Agriculture: *Exorista lavarum, E. segregata, E. rossica* (Diptera: Tachinidae), and *Rogas indiscretus* (Hymenoptera: Braconidae) (Metterhouse, 1974). It is hoped that these parasites may be colonized in gypsy moth-infested areas of the Northeast.

This tiny parasitic fly, *Palexorista sp.,* is shown depositing her eggs through the protective hairlike structures of the gypsy moth caterpillar into its body wall. *Courtesy: USDA*

The parasites *A. melanoscelus* and *C. concinnata* have become established in Canada. Alternate hosts for many of the established gypsy moth parasites in the Northeast included the variable oak leaf caterpillar, *Heterocampa manteo;* the red-humped oak worm, *Symmerista canicosta;* the orange-striped oak worm, *Anisota senatoria;* the pink-striped oak worm, *A. virginiens;* the fall webworm, *Hypantria cunea;* and other oak leaf rollers (Quimby, 1973).

Presently, parasitic material and information are being obtained from France, Germany, Spain, India, and Japan; much of the material received is in diapause. In France the nun moth, *Lymantria monacha,* a close relative of the gypsy moth, has gradually replaced the gypsy moth in northern

France. Since these two moths are close relatives and parasites of one commonly attack the other, investigators are focusing on parasites of the nun moth for possible use in gypsy moth-control programs in North America (Sailer, 1973). Parasites of the nun and gypsy moths obtained from northern France and Germany, if successful, might be better suited to the colder climatic conditons of the northeastern United States than those parasites from southern France and Spain. The success of these parasites would be of great value to the biological control program since several important gypsy moth parasites overseas have failed to become established in New England, mainly as the result of low winter temperatures.

In all, there are about 20 parasites in the Northeast that attack and control the egg, larval, or pupal stage of the gypsy moth to come degree, but none are available commercially. In fact, many of the introduced parasites and predators are nearly as effective in North America as they are in Central Europe. These parasites are being reared and released by federal and state agencies throughout the Northeast. The New Jersey and Pennsylvania Departments of Agriculture have been increasing and broadening their biological control programs against the gypsy moth within the past five years.

Released parasites increase in cycles following gypsy moth population increases. Therefore, there is a time lag in effectiveness before any permanent control of the gypsy moth can be observed. This indicates the necessity for early detection of the insect in newly infested areas, if biological control is to be utilized.

Present objectives of parasite programs include the following: (1) to reduce the periodic outbreaks of the insect, (2) to reduce the insect to the level of a native pest insect, (3) to colonize new species of parasites in order to increase biological control of the gypsy moth, and (4) to colonize newly infested areas with imported and established parasites. Several programs are focusing on the discovery of which parasites are most effective in controlling the gypsy moth at low population levels. On some occasions, introduced parasites may become more effective in the new region than in their normal habitat; careful investigation is necessary to find such parasites. Often pupal-collecting programs are organized for July through December. Collected pupae are incubated; then any parasites recovered from the pupae are held until maturity and then released into areas where they have not yet become established.

Parasitic insects may be released in a series of steps to attack the gypsy moth. From July through December eggs may be attacked before they hatch by the Japanese wasp, *Ooencyrtus kuwanae*. Caterpillars from surviving eggs may then be attacked by *Apanteles melanoscelus*. Following the release of *A. melanoscelus* an additional parasitic wasp, *Brachymeria intermedia*, may be released to attack pupae of the gypsy moth. Furthermore, this series of released parasites may be supplemented throughout the summer by the use of insect disease agents, such as *Bacillus thuringiensis* and the nuclear polyhedrosis virus, and by the use of insect and mammalian predators.

There are many phases of a biological control program. Initial emphasis is placed on the screening and rearing of introduced parasites and predaceous insects whose primary host is the gypsy moth. This task has been made easier with the development of a means for continuous laboratory rearing of the gypsy moth by the Plant Protection Division's Method's Laboratory, Otis Air Force Base. The gypsy moth and a closely allied moth species, *Lymantria obfuscata*, along with the greater wax moth, *Galleria mellonella*, can be used to mass-rear gypsy moth parasites. The salt marsh caterpillar, *Estigmena acrea*, also can be used to mass-rear gypsy moth parasites.

For a long time, gypsy moths were mass-reared on natural foods for parasite programs and experimental studies (Forbush and Fernald, 1896; Goldschmidt, 1934; and Maksimovic, 1958). However, starting in 1966 and continuing today, gypsy moths are mass-reared on an artificial diet (Leonard and Doane, 1966). Mass rearing of the gypsy moth has become standardized: Eggs are removed from the mass and depilated, and their surfaces are disinfested with 0.5% sodium hypochlorite for 30 to 45 minutes to remove microbial contaminations; then they are rinsed twice in sterile water (Leonard, 1970; O'Dell and Rollinson, 1966). If the eggs are to be stored, they are packaged and incubated at $10°C$. The manner in which the eggs are stored affects the percentage that hatch. Eggs packaged in an immobile condition in taut saran packets closely resemble an undisturbed egg mass and pield a significantly higher percentage of hatching than under other storage conditions (Marx, 1974b).

After hatching, the larvae may be placed in individual petri dishes or grouped in colonies and fed on an artificial diet consisting mainly of modified wheat germ. Temperature and relative humidity must be maintained at fairly constant levels. Larvae are incubated at a temperature between $+23°C$ to $+26°C$ and at a relative humidity between 40% to 60%. The larvae must also be subjected to photoperiods of 14 to 16 hours per 24-hour day; this time period must be checked at regular intervals. Through the process of mass rearing, eggs and larvae may be evaluated quantitatively for eggs/mass counts, percent of parasitism, and percent of nonviability. Gypsy moth rearing is difficult, since there is a heavy dependence on field-collected eggs, which often carry disease.

In addition, the gypsy moth experiences normally high mortality rates during rearing periods. Laboratory mass-rearing of the gypsy moth is a combined effort of basic and applied research. Three areas of gypsy moth research dependent on laboratory-reared gypsy moths are parasite control programs, pheromone research, and the sterile-male technique.

Many reared parasites are released in areas where gypsy moth infestations are new, in order to reduce any potential outbreaks. Those parasites which failed in New England are being reared again in an attempt to establish them in the more southerly infested areas of the Northeast since environmental factors in these areas may be more desirable for their establishment. Foreign and domestic searches are conducted to investigate additional parasites. There appear to be at least 24 species of gypsy moth parasites or predators in Japan alone that have not yet been introduced in North America (Metterhouse, 1973). There also are two important Coleopteran egg predators in Europe, *Dermestes erichosi* and *D. lardarius* (Stefanouv and Keremiokhiev, 1961). In order to obtain information on the effectiveness of many gypsy moth parasites, the USFS utilizes radiography on such parasites as *Blepharipa scutellata* to determine development and mortality within the gypsy moth host (Marx, 1974a).

Before their release, all introduced parasites and predators are screened adequately through a federal quarantine program, and therefore they present no environmental hazards. Whenever parasites or predators are to be introduced into a new habitat, they must be studied to ensure that they themselves do not become a problem. If not screened, they may present such potential problems as transmitters and vectors of diseases and parasites of valuable insects or other parasites (hyperparasitism). In 1972, 5,629 parasitic flies and wasps were sent from Sèvres, France, to the Agriculture Research Service insect quarantine laboratory at Moorestown, New Jersey (Agr. Res., 1973). Before shipment these parasites were shown to be harmless to man, plants, animals, and insects other than the gypsy moth.

Parasites must be released in large numbers and at regular intervals until they become established. In 1971 more than four million parasites were released by federal and state agencies in ten states. The New Jersey Department of Agriculture reared and released over 225,000 parasites in 1973 (Metterhouse, 1974); in 1972 the Pennsylvania Department of Agriculture reared and released over 70,000 parasites (Nichols, 1972). From 1972 to 1973 over 1,447,000 parasites were released throughout the eastern United States (Marx, 1972c); and in 1974 almost 1,100,000 parasites were reared in Pennsylvania alone for further release programs (Marx, 1974b). In eastern West Virginia 20,000 *B. intermedia* were released during 1973 (Marx, 1973b).

Field evaluations of parasitic-control programs consist of collections of efficiency data on parasitism and determination of parasite establishments. This is done by collecting eggs, larvae, and pupae in the field that were exposed to released parasites and then examining them for signs of parasitism. Vectoring and dissemination of diseases by parasites also are under investigation.

In order to determine which parasites should be imported, field research areas throughout the Northeast are monitored to reveal which parasites are attacking the gypsy moth. In addition, parasite effectiveness in terms of numbers of female gypsy moths that survive to produce the next generation is determined by a life-table approach (Marx, 1972b).

Biological control methods are selective and expensive means of past management. By 1971 this form of pest management cost $360,000 annually to develop; and William Gillespie, former chairman of the National Gypsy Moth Advisory Council, estimated that $875,000 more would be required annually from 1972 to 1977 if favorable results from biological control programs were to be obtained.

Parasites and predators are beginning to illustrate significant effectiveness in controlling the gypsy moth. Recently they have held the insect to manageable levels in the New York Champlain Valley, in Massachusetts (Marx, 1972b); and in many parts of the generally infested areas of Connecticut the white-footed mouse, *P. leucopus,* is given credit for maintaining the gypsy moth at low population levels (Marx, 1973a).

Biological control programs presently appear to offer the only known means of reducing the gypsy moth to the same status at the native forest insects. The choice of biological methods varies from area to area depending on the forest type, land usage, and characteristics of the gypsy moth infestation itself; but the insect remains vulnerable to the atacking parasites and predators. Perhaps it can be siginficantly controlled as this approach becomes more and more effective.

A lost of over 300 selected references pertaining to gypsy moth parasites and predators has been prepared by Hanson and Reardon (1973).

Table 11 - 13

MAJOR ESTABLISHED OR MASS-REARED INSECT PARASITES AND PREDATORS OF THE GYPSY MOTH IN NORTHEASTERN UNITED STATES

Coleoptera
    Carabidae

Calosoma frigidum	larval predator
Calosoma scrutator	larval predator
Calosoma sycophanta	larval predator
Calosoma wilcoxi	larval predator

    Tenebrionidae

Tarpela micans	larval predator

Diptera
    Tachinidae

Blepharipa scutellata	larval parasite
Compsilura concinnata	larval parasite
Exorista larvarium	larval parasite
Exorista segregata	larval parasite
Palexorista inconspicus	larval parasite
Parasetigena agilis	larval parasite
Lespesia sp.	larval parasite

Hemiptera
  Pentitomidae

Apatelicus cynicus	larval scavenger
Podisus maculiventris	larval scavenger
Podisus placidus	larval scavenger

Hymenoptera
  Braconidae

Apanteles lipardis	larval parasite
Apanteles melanoscelus	larval parasite
Apanteles porthetriae	larval parasite
Meteorus pulchriconrnis	larval parasite
Rogas indiscretus	larval parasite

  Calcidae

Brachymeria intermedia	pupal parasite

  Encyrtidae

Ooencrytus kuwanae	egg parasite

  Ichneumonidae

Campoples sp.	larval parasite
Coccygomimus instigator	pupal parasite
Coccygomimus pedalis	pupal parasite
Coccygomimus turionella	pupal parasite
Cratichneumon w-album	larval parasite
Itoplectis conquisitor	pupal parasite
Phobocampe disparis	larval parasite
Scambus sp.	larval parasite
Theronis atalantae	pupal parasite

12

Microbial Control

Because of recent restrictions on pesticides, microbial control of insect pests, particularly the gypsy moth, has received much attention. Microbial control, a form of biological control, is referred to as the use of microorganisms by man to control pest species. There are several viruses, rickettsiae, bacteria, nematodes, protozoa, and fungi that are lethal to a number of lepidopteran larvae, including the gypsy moth. Of all these microorganisms *Bacillus thuringiensis* Berliner (Bt) and *Borrelinavirus reprimens* have received the most attention for use as control agents against the gypsy moth.

There are four basic routes by which microorganisms invade their hosts. These routes are oral infection, integumentary or tracheal infection, parenteral invasion, and transovarian invasion. Oral infection occurs when a microbe enters a host through the digestive tract via ingestion. Integumentary or tracheal infection occurs through the intact integument or tracheal system and is the most common route of invasion by many microorganisms, along with oral infection. Parenteral invasion occurs when a microbe invades the body of a host when its integument is broken. In transovarian invasion the transfer of pathogens occurs during ovipositing from the environment or the body wall of the adult female insect to the egg shells. In transovarian invasion, pathogens may also be transferred from the ovary to the eggs before ovipositing.

Microbial invasion or the dissemination of micro-organisms is enhanced by three methods. In the first, when an insect dies of a microbial disease, its body disintegrates and the microorganisms contained in the cadaver are dispersed by the wind to other insects and host plants. In the second, microorganisms in the gut of an infected insect are distributed in the feces or frass of the insect. Finally, insects infected by microbes can be eaten by predators, which then spread the microorganisms in their feces.

BACILLUS THURINGIENSIS BERLINER

E. Berliner first described the existence of a parasporal body in sporulated cells of *Bacillus thuringiensis* in 1915 (Abbott Lab., 1971). However, the potential role of this microbe for insect control was not developed at that time. Study of the organism for use in insect control began at the University of California in 1948, but it was not until 1953 when C. L. Hannay demonstrated the significance of this parasporal body as a toxic proteinaceous crystal (Abbott Lab., 1971). Research was undertaken then, and the bacteria have since been explored extensively as insect control agents.

Bacillus thuringiensis Berliner is a naturally occurring, gram positive, spore-forming bacterium, Bacillaceae, that is pathogenic to gypsy moth larvae when present in sufficient numbers. There are several strains of the bacterium that are specifically lethal to many lepidopterans. This specificity is due to the alkaline condition required by the bacteria to exert their lethal effect. This alkaline environment is found in the gut of most lepidopterans, while the gut environments of other forms of animal life are neutral to acid. Once inside the gut of the larvae, only a few of the spore-forming bacteria prove pathogenic to the insect. This is because most bacteria cannot pass through the wall of the insect's gut. However, some of the consumed bacteria produce toxins that cause buccal and gut paralysis, accompanied by extensive changes in the gut epithelium permitting invasion of the insect's body cavity. Cessation of feeding occurs when the walls of the midgut are disrupted. Once the bacterial-formed spores enter the blood, they generate into rods. In this new form the bacteria are lethal to the larvae.

These pathogenic bacteria produce four toxic substances: a parasporal body, which is the chief endotoxin; a dialyzable molecule, phospholipase C or lecithinase, which is the chief exotoxin; and an additional phospholipase. The parasporal body is a proteinaceous crystalline material produced inside the bacterial cells during sporulation that is capable of palayzing the gut of the insect. The dialyzable molecule is small, heat stable, and soluble; it is produced in the fluids of the insect's body cavity. This molecule is thought to be the agent responsible for damage caused to ATP metabolism of the host; it is capable of killing the insect although its exact means is not yet known. Phospholipase C, an enzyme produced by the growing bacteria, serves to break down essential phospholipids in host cell membranes, causing hemolysis. The additional phospholipase also affects life-supporting phospholipids of the insect. Larvae infected with pathogenic bacteria cease to feed, become sluggish, and die within three or four days of infection. The DNA of the bacteria is circular, with a molecular weight of 50×10^9 (Marx, 1974b).

The biological insecticide Bt can be used with reasonable safety near ponds, lakes, and streams, and it is harmless to man, wildlife, beneficial insects, and predators and parasites of the gypsy moth. Further, no evidence of foliar burn has ever been observed with the use of Bt. There is no food chain magnification such as those that occur with many chemical insecticides, nor is there any known negative effect to the ecosystem.

Bacillus thuringiensis can be propagated or fermented on an artificial medium of fairly standard composition. This medium includes a salt mixture, a protein source, sometimes a vitamin, and a growth-factor source such as yeast. Bt can be produced in fermenters of varying sizes. Organisms are obtained from the fermenters either by vacuum-drying the medium or by centrifugation to a slurry, which is eventually dried or stabilized for storage. Production of an effective bacterial preparation is dependent upon the presence of active toxins and visible spores.

There are three U.S. companies producing commercial preparations of *B. thuringiensi* insecticide. These are Sandoz, Nutrilite Products (material trademarked Biotrol XK), and Abbott Laboratories (material trademarked Dipel). These preparations are registered for use by the EPA against the gypsy moth. These various commercially named formulations provide roughly the same results at adjusted rates (Dunbar and Kaya, 1972). Several are especially formulated for high-pressure, high-volume aerial application and can be mixed with water and sprayed on trees in the same manner as chemical insecticides. In addition to water suspensions, Bt can be applied in oil-water emulsions and in the form of bacterial powders with clay or as granular formulations with clay. The adjuvants help to ensure adherence of the spray deposits to the foliage. The choice of application material depends on the volume of spray and type of application. The effectiveness of the Bt spray can be enhanced with the creation of an acidic reaction during spraying operations. The addition of boric acid to the *Bacillus* mixture increases larval mortality and substantially increases pathogenicity of sublethal doses (Doane and Wallis, 1964). Bt has even been mixed with molasses to produce control measures comparable to some chemical insecticides (Marx, 1972b).

Bt is not effective as a contact insecticide since the microbial agent must be ingested in order for it to kill. Therefore, proper timing of the application and coverage of foliage is necessary to ensure larval consumption of a lethal dose of the bacteria for effective control. Good foilage protection is possible if egg-mass counts are under 500 per acre (Marx, 1974b). Spraying is correlated with foliage development, usually when leaves have expanded 30 percent. This timing places the bacteria on the foliage when the larvae are in their second and third instars. If the eggs are late in hatching, Bt is applied when the leaves have expanded 50 percent. In heavy infestations a second spraying usually is applied seven to ten days after the first.

Spraying generally remains effective for one month; however, this period may be shortened considerably by heavy rainfall following application of the bacteria. Persistence of the microbe is important since the eggs may hatch over a four-week period. Bt can be applied with ground or aerial spraying equipment. When applied from the air, Bt produces variable results and generally fair control of larvae, while ground applications appear to yield better and more effective results.

Since the bacteria must be ingested, the distribution and amount of Bt deposit on the foliage are important factors when the microbial agent is applied. The amount of spray material required to obtain effective coverage varies with the height and crown-type of the trees and with the type of spraying equipment used. Cantwell and co-workers (1961) reported that a spore concentration of 1.4×10^9/ml solution was necessary for effective control. Doane (1966) noted that ground aplication of Thuricide 90 TS at 10 and 20 pints in 10 gallons of water effected significant gypsy moth mortality. Bt is usually applied at a rate of eight billion International

Units per acre (Marx, 1974a). Lewis (1973b) reported that six billion International Units per acre yielded good foliage protection and that a single application of eight billion International Units was not as effective as the double application of four and six billion International Units. With increased concentration of the bacteria in the spray, there is an increased percentage in the larval mortality rate (Yendol et al., 1973).

In many cases Bt used against gypsy moth infestations yielded generally good foliage protection but poor population reduction; this is due to the feeding inhibition effect of Bt. Larvae may consume sufficient bacterial material to inhibit feeding, but they may not consume a lethal dose. Therefore they can survive and produce generations capable of defoliation the following year. Good egg-mass reduction but high foliage damage also may occur because of the delayed effect of Bt on gypsy moth larvae (Marx, 1974b). In most investigations of the effectiveness of *Bacillus thuringiensis* against the gypsy moth, there has been a need for high concentration of Bt spores to achieve significant mortality within the population. Definite proof of the protection of forest and shade trees by Bt against gypsy moth defoliation has been demonstrated by the Connecticut Agricultural Experiment Station, New Haven. However, adequate reduction of gypsy moth populations can be achieved only when egg mass counts are under 100 per acre (Pa. Dept. of Env. Res. 1976).

Bacillus thuringiensis also can be used along with other pesticides if these materials are used within a few hours of mixing. A number of acaricides, insecticides, fungicides, adhesives, and wetting agents are compatible with Bt and can be mixed in the same tank with the bacteria.

Like other insecticides, Bt also has some disadvantages. Even with several applications of spray material some defoliation can be experienced, and the nuisance problem of migrating caterpillars still exists. Also, the cost of this biological insecticide, estimated at 15 dollars per acre, is high when compared to other materials. Finally, BtB is made inactive by sunlight and is ineffective when washed off the foliage by rain.

Not all species of lepidopterans are equally susceptible to Bt. The rate of Bt per acre required to control some species, such as the cankerworm, *Alsophila pometaria,* is at least one half the rate required to control the gypsy moth. Bt also has been used in the control efforts of other pest insects such as the cabbage looper, *Trichoplusia ni;* the imported cabbagewgorm, *Pieris rapae;* the tobacco budworm, *Heliothis virescens;* the tent caterpillar, *Malacosoma americanua;* the forest tent caterpillar, *Malacosoma disstria;* and the alfalfa caterpillar, *Colias eurytheme.*

The objective of using *Bacillus thuringiensis* is to control 80% to 90% of gypsy moth larvae within an infestation and thereby prevent massive defoliation. Hopefully, the remaining 10% to 20% of the larvae would serve as hosts for beneficial parasites and predators, which could regulate the surviving pest population. If this objective can be accomplished, *Bacillus thuringiensis* would provide an effective means of control for the gypsy moth.

Bt presently is undergoing extensive testing in central Pennsylvania to determine its ability to prevent oak and white pine mortality from gypsy moth damage and to slow the insect's rate of spread from the Northeast (Marx, 1974a).

Table 12 - 14

MATERIALS COMPATIBLE WITH *Bacillus thuringiensis*

Aramite	Dodine	Parathion
Bidrin	Dyrene	Phaltan
Bordeaux mixture	Endosulfan	Phosdrin
Captan	Endrin	Phosphamidon
Carbaryl	Folpet	Phygon

Cygon
Daconil
DDD
DDT
Diazinon
Dibrom
Dichlone
Dieldrin
Difolatan
Dimethoate
Dithane M-22
Dithane M-45
Dithane Z-78

Guthion
Karathane
Kelthane
Magnetic 6
Flowable Sulfur
Malathion
Maleic hydrazide
Maneb
Meta-Systox R
Methyl parathion
Methyl Trithion
MH-30
Naled

Pyrethrum
Rotenone
Ryania
Sorba Spray
Strobane
Sulfur
TDE
Tedion
TEPP
Thiodan
Trichlorofon
Toxaphane
Trithion

(Courtesy: Sandox Corp., after Inter.
Minerals and Chem. Corp., 1971)

BORRELINAVIRUS REPRIMENS—"WILT" DISEASE

Under crowded conditions of larval development, large numbers of gypsy moths die of a viral disease known as "wilt" disease or nuclear polyhedrosis. The causative agent for this disease is a nuclear-polyhedrosis virus, *Borrelinavirus reprimens,* always present in gypsy moth colonies in a latent form. This virus is named for its presence in the nucleus of host cells. Under natural conditions of stress such as starvation or high population densities, the virus reaches epizootic levels. It is this epizootic condition of the virus that indicates the end of a dense larval population. Because of the regulatory action of the virus upon the gypsy moth, it appears as a potential biological weapon. The action of the virus is similar to that of Bt, that is, it is specific only to lepidopterans. The polyhedral virus disease of the larvae is considered to be the most important natural agent causing gypsy moth populations to collapse.

The virus first appeared in the United States in 1907 in Massachusetts and currently inhabits most infested areas. In 1951 laboratory and field tests conducted by the Forest

Insect Laboratory, New Haven, revealed that the polyhedra could initiate epizootic conditions and might provide some measure of control against thse gypsy moth when applied as a spray on the host plants (Rollinson et al., 1965). In 1971 at the Agricultural Experiment Station, Department of Entomology, Pesticide Research Laboratory at Pennsylvania State University, the virus was isolated and purified from diseased larvae found in the Kunkletown area of Monroe County, Pennsylvania (Hamlen, 1972).

Several sources of the virus exist. Viral sources from Japan, Yugoslavia, and other countries have been tested against the gypsy moth in North America. The results of these tests indicate that all foreign sources tested are significantly less potent against gypsy moth populations than any native polyhedrosis virus (Marx, 1970a). Within the United States differences exist in the effectiveness of the same virus preparation against gypsy moth populations in different geographic areas. In one test highly purified, native nuclear polyhedrosis virus was six times more effective against Michigan larvae than against Massachusetts larvae (Marx, 1970a).

The virus particle or virion is enclosed inside a protein crystal, the polyhedron. This protein coat provides the virion with protection, enabling it to retain its infectivity for a long period of time, which may extend over several years. The polyhedra are irregular in shape, ranging in diameter from 6 to 15 microns, while the virus particle measures 18 x 280 millimicrons, visible only under an electron microscope (Hamlen, 1972).

The usual invasion route of the virus is by ingestion or through the tracheal system. Once entrance is achieved, the virus attacks the cell nucleus or cuticle, fat body, blood, or tracheal cells.

The virus also attacks the polyribosome system of the infected cell and begins producing additional viral material, resulting in a steady decrease in the polyribosome content of the host cell (App and Granados, 1972). In the early stages of infection, the larvae become inactive, sluggish, and swollen in size, and they cease to feed; this is accompanied by little growth or weight increase in infected larvae. The larvae soon begin to assume an oily or a greasy appearance. As the infection progresses, the cuticle becomes fragile and brittle and ruptures easily, releasing the liquid body contents. The exuding fluids contain much infectious viral material, which then spreads the "wilt" disease to other larvae. In later stages of the disease, the infected larvae hang their heads downward in a wilted appearance, suggesting the term *wilt disease*. As the infection goes into its final phase, the number of proteins in the blood also increases due to protolysis.

Increased incidence of the virus occurs under certain circumstances such as high, prolonged humidity, periods of stress, crowding, and starvation. At the peak of larval activity a sharp increase in diseased larvae was reported in a large infestation in Massachusetts during a period of hot and humid nights (Marx, 1974b). Cold periods just before egg hatching may greatly increase the percentage of larvae that succumb to the disease in the first instar (Doane, 1969) Diet also may affect the incidence and spread of the virus. Larvae that feed on the large-tooth aspen, *Populus grandidentata*, were observed to have a higher disease incidence than those which did not feed on the aspen foliage (Mosher, 1915; Bess, 1961).

The viral material is easily obtained from naturally infected larvae and can be laboratory-reared in larvae to increase its quantity. This is accomplished by adding purified polyhedra to a synthetic diet, which is then fed to the larvae. The virus can reproduce only within a polyhedron in living gypsy moth larvae. Four gypsy moth cell lines have been developed by the Agricultural Research Service for viral research and control work. Three of these lines support full replication of the gypsy moth virus to polyhedra reproduction, and one produces only the virion (Marx, 1974b).

The virus can be produced in purified preparations by the use of several simple laboratory steps. First, the virus is isolated from infected ground-up larvae in an aqueous suspension. In order that the larval debris be removed, the suspension is passed through cheesecloth and then centrifuged at low speed. Next, by spinning the supernatant through a series of sucrose solutions, the polyhedra are isolated at a density level corresponding to a particular sucrose solution. The trapped polyhedra-sucrose solution is removed, diluted with water, and centrifuged again. Finally, the supernatant is removed and pellets of polyhedra are removed in a small amount of acetone and allowed to dry. The polyhedra are refrigerated if they are not immediately needed.

Hamlen (1972) has suggested a lethal concentration of the virus to be 4.11×10^3 polyhedra per milliliter of diet, while Rollinson and co-workers (1965) required a water-spray concentration of 2.7×10^8 polyhedra per milliliter of water for effective gypsy moth control. The concentration of polyhedra necessary for control is dependent upon the degree of infestation in an area. However, the mortality rate of the gypsy moth increases with increasing concentrations of the virus.

Before the virus can be utilized as a control agent, its action in the environment of the gypsy moth must be understood. Hamlen (1972) has suggested several questions that can be used as guidelines for the investigation of the virus: (1) Why does the virus appear active only under crowded or dense larval populations? (2) What influence do environ-

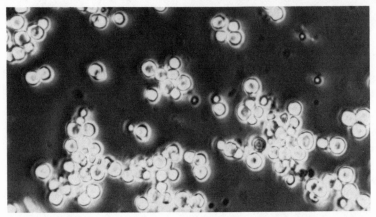

Polyhedra isolated from virus-killed larvae. *Courtesy: R. A. Hamlen*

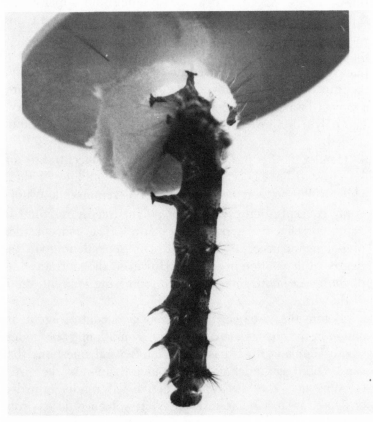

Gypsy moth larva infected with nuclear polyhedrosis virus. Larvae affected with a late stage of the virus show a characteristic pattern of hanging the head downward. *Courtesy: R. A. Hamlen*

mental factors have on the incidence of the disease? (3) How is the disease spread through a larval population? (4) What transovarian properties does the virus have? (5) What relationship is there between virus concentration and short- and long-term control efforts?

Present plans with the polyhedrosis virus include formulation studies, viral propagation studies, biochemical and biophysical characterizations, safety studies, field and virus vector studies, and enchancement activity studies (Lewis, 1973a). In addition to these plans, the USFS is attempting to develop a virus-free strain of the gypsy moth for comparison purposes (Marx, 1974b). The USFS also is modifying its present procedure, the Bergold procedure, for releasing the virus from occluding polyhedra. Future plans include the introduction of specific and sensitive systems for the *in vivo* and *in vitro* detection of the virus.

The virus appears to have an advantage over Bt. Once introduced into a gypsy moth population, the virus could be spread like a highly contagious disease by infected caterpillars to healthy ones. This would make it possible to use inoculated larvae as a vector of the disease, thereby reducing the amount of virus necessary for a disease outbreak. Unfortunately, the virus does not appear to suppress a significant number of larvae until gypsy moth populations are very dense and severe defoliation has already occurred. At the present time, the low profit potential of the virus and its need for further testing make it useful only as a potential control weapon.

13

Gypsy Moth Sex Attractant

An increased interest in the use of insect attractants for pest control also has resulted from the present trend toward minimizing pesticide usage. Sex attractants are used for many purposes : as lures in traps to sample insect populations; to determine the relative densities of insect populations; to follow insect movement and dispersal patterns; and to control and manage pest populations.

Insects can influence other members of their species by means of chemical stimuli. These chemicals used as a system of communication within an insect community are called pheromones. Sex pheromones play an important role in the mating behavior of the gypsy moth. The pheromones are secreted to the outside environment by the female and received by the male. Acting on the sensory organs of the male, the pheromones serve mainly as an attractant causing a locomotory response from a distance and possibly as an aphrodisiac stimulating copulation.

The majority of the pheromone released is deposited on the glandular surface of the female. From here it is dissipated in the air by evaporation and by fanning of the female's wings. The attractant then is subjected to the movement patterns of the air.

The pheromone concentration in the air is the highest

in the vicinity of the female; away from the female a pheromone gradient is established in a general ellipsoidal shape. The dimensions of the ellipsoid-shaped gradient and the concentration of the pheromone within its boundaries depend on the evaporation rate of the pheromone and environmental factors such as wind speed, humidity, and temperature. Males are attracted to pheromone-extract baited traps with increasing relative humidity, and male flight responses to the female increase with increasing temperatures from $+21°C$ to $+33°C$, while no response occurs at temperatures below $+20°$ (Collins and Potts, 1932).

In order to detect the sex attractant, the male's antennae are pectinate in structure and highly developed, responding specifically to the species pheromone. The antennae are so highly selective of the pheromone that its detection cannot be masked by the presence of strong-smelling substances. The male, capable of detecting a female as far as seven miles away, responds to four millionths of a billionth of an ounce of sexual attractant (Leonard, J., 1972).

Through the use of pheromone, a design for behavioral control programs is being formulated to combat the gypsy moth. These control programs are directed against the large numbers of pheromone receivers and are designed to survey, confuse, and trap the male gypsy moths. Confusing the males in their attempts to find mates and reducing the number of males through trapping programs result in a decreased number of matings in a given population. Used in this manner the pheromone prevents a large number of females from laying fertile eggs; thus, the end result of such behavioral control programs is the reduction of fecundity and population density of the species in a treated area. The use of a synthetic lure in place of the natural pheromone also has many advantages, including the operation of traps during periods of time when the female is not available, the use of a more readily adjustable rate of pheromone release, and the reduction of operational costs.

In 1893 A. H. Kirland first attempted to capitalize on the attraction of females for males in gypsy moth control work (Farsky, 1938). It was found later that a crude extract of the last two abdominal segments, "tips," of the virgin female contained a sex attractant capable of luring the male. This extract was used in survey traps to detect infestations of the insect until the early 1970s. Each year hundreds of thousands of females were required to furnish enough lure for such survey programs. These programs consisted of approximately 60,000 traps and operated at a cost of about $30,000 annually during the 1950s and 1960s. Female pupae were collected in the field from Spain and France. After the female had emerged for a 24-hour period, the last two abdominal segments were removed and placed in benzene. The tips were extracted with benzene, and the solution was concentrated. The resulting pheromone extract was hydrogenated, and the survey traps were baited with the equivalent of ten female tips. Because of the great number of female moths needed and the high cost of such survey operations each year, experimentation to produce a synthetic attractant to replace the natural lure was undertaken in the early 1920s at Harvard University. This research work was continued by the USDA in 1940.

Early chemical studies were undertaken by H. L. Haller and co-workers (1944); however, it was one of Haller's associates, F. Acree, Jr. (1953 and 1954), who determined that the sex attractant was a lipid constituent of the unsaponifiable fraction of a benzene extract from the last two abdominal segments of virgin females.

In 1960 Jacobson and co-workers isolated the supposed sex attractant "gyptol," *cis*-7-hexadecene-1, 10-diol 12-acetate (6), which has a molecular weight of 298. This compound was isolated from an

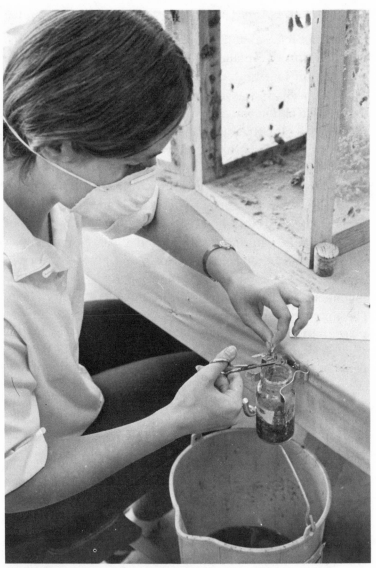

Before the synthesis of disparlure, the sex pheromone for gypsy moth control work had to be extracted from the female adult. The snipping of rear sections from the female was slow work and required protective face masks to prevent tiny insect hairs and scales from irritating the laboratory worker. *Courtesy: USDA*

$$CH_3(CH_2)_5CHCH_2CH=CH(CH_2)_5CH_2OH$$
$$OCCH_3$$
$$O$$

Gyptol (6)

unsaponifiable fraction of 200,000 female tips according to the chromatographic techniques of Acree. Gyptol was synthesized in both liquid and granular forms. However, the compound was lacking in activity during field tests when distributed by aircraft over Rattlesnake Island in Lake Winnipisaukee, New Hampshire. Field observations revealed that gyptol was only weakly attractive and could be masked by chemicals produced during the spraying operations (Jacobson, 1965).

In 1962 Jacobson and Jones synthesized a more potent homologue to gyptol, differing only in having two more carbon atoms with bonded hydrogen on its chain length. This synthetic sex attractant was prepared from ricinoleyl alcohol and was called "gyplure," *cis*-9-octadecene-1, 12-diol acetate (7), which has a molecular weight of 326. Although gyplure was more potent than gyptol and less expensive to synthesize than preparing the natural

$$CH_3(CH_2)_5CHCH_2=CH(CH_2)_7CH_2OH$$
$$OCCH_3$$
$$O$$

Gyplure (7)

sex lure, it was still 100-fold less active than the natural pheromone when tested in the field (Jacobson and Jones, 1962). Gyplure also was subject to contaminant masking by ricinoleyl alcohol and to inactivation by the formation of the *trans* structure of gyplure. As little as three to seven percent ricinoleyl alcohol was sufficient contaminant to effect little or no biological activity (Waters and Jacobson, 1965; and Jacobson, 1966). With only two carbon atoms of difference between gyplure and the natural pheromone, there was almost complete loss of biological activity.

In 1967 the hydrocarbon fraction of an extract of 385,000 tips collected in Spain was separated and treated A highly active epoxy compound was isolated, which was eventually used for the identification of the natural pheromone. In 1969 the gypsy moth sex attractant was extracted by Bierl and co-workers from 78,000 tips at the Pesticide Chemical Research Branch of the USDA, Entomology Research Division. The pheromone was identified as *cis*-7, 8-epoxy-2-methyloctadecane (8), which has a molecular weight of 282. This isolation and identification

$$(CH_3)_2CH(CH_2)_4\overset{H}{\underset{\underset{O}{\diagup}}{C}}\text{-}\overset{H}{\underset{}{C}}(CH_2)_9CH_3$$

Disparlure (8)

of the pheromone enabled Bierl to synthesize the gypsy moth sex attractant for the USDA. The synthetic sex attractant is known as disparlure.

Males react to as little as 10ug of disparlure (Beroza, 1971); 1 ug is about equal to the attractiveness of a virgin female (Beroza et al., 1971b). The first preparations of disparlure resulted in a 30-gram unit sufficient to bait traps well beyond all practical needs and made it available in unlimited amounts. The cost of disparlure is estimated to be thirty cents per gram. The specificity of the sex attractant is evident from the great difference in activity that results from slight variations in molecular configuration. The high potency and narrow specificity of disparlure are due to its high molecular weight.

Since the adult flight season may last from four to six weeks in any location, it is important that disparlure retain its effectiveness to disrupt mating or attract males over that length of time. To regulate the volatization and to prolong the effectiveness of the synthetic pheromone, it is mixed with "keepers." The keeper most often used with disparlure is trioctanoin, a large, nonvolatile, saturated liquid gylceride with a molecular weight of 471. The disparlure-trioctanoin combination is far superior to the natural moth extract in both intensity of attraction and persistence. Traps containing 1 ng of disparlure plus 5 mg of trioctanoin are capable of catching moths after field exposure for 12 weeks (Beroza and Knipling, 1972). Trioctanoin also has been observed to enhance the attractiveness of disparlure, possibly with the production of octanoic acid by hydrolytic cleavage (9). Because of the effectivenes and longevity of the disparlure-trioctanoin formulation, traps need not be rebaited during

$$
\begin{array}{ll}
\text{H}_2\text{C–O–}\overset{\displaystyle \text{O}}{\overset{\|}{\text{C}}}\text{–(CH}_2)_6\text{–CH}_3 & \text{H}_3\text{C} \\
\quad | & \quad | \\
\text{HC–O–}\underset{\displaystyle \text{O}}{\text{C}}\text{–(CH}_2)_6\text{–CH}_3 \;\rightarrow\; & \text{HC–O–}\overset{\displaystyle}{\underset{\displaystyle \text{O}}{\text{C}}}\text{–(CH}_2)_6\text{–CH}_3 \\
\quad | \quad & \quad | \\
\text{H}_2\text{C–O–}\underset{\displaystyle \text{O}}{\text{C}}\text{–(CH}_2)_6\text{–CH}_3 & \text{H}_3\text{C–O–}\underset{\displaystyle \text{O}}{\text{C}}\text{–(CH}_2)_6\text{–CH}_3 \\
& + \\
& \text{HOOC–(CH}_2)_6\text{–CH}_3 \\
& \text{octanoic acid}
\end{array}
$$

Trioctanoin cleavage producing octanoic acid (9)

the trapping season. This alone has saved government agencies over $50,000 yearly. Other fairly effective keepers used with disparlure are 1-octadecanol, methyl nonadecyl-ketone, 1-bromocadecane, methyl stearate, eicosane, and palmitic acid.

Disparlure can be used to survey and detect the occurrence and relative population density of the gypsy moth and also to contain and manipulate field populations of the insect at low densities. Positive, straight-line relationships on log-log plots between males captured and attractant per trap suggests that disparlure would be effective in estimating gypsy moth population sizes in light infestations. At high population densities, however, the traps would become saturated with moths and an estimate of the population size would be difficult.

Disparlure can be used in a variety of ways to contain and manipulate field populations of the insect. Cameron (1973) has suggested four uses of disparlure for such operations: (1) the eradication of spot infestations located outside any generally infested area, (2) the clean-up of residual populations after other control agents have reduced the insect's population density in an infested area, (3) the general restriction of infestations to the Northeast and reduction of the generally infested area, and (4) the establishment of a barrier zone wide enough to prevent the formation of new infestations from wind-dispersed larvae. Many believe it is possible to slow and perhaps halt the natural spread of the gypsy moth by constructing a 100-mile-wide buffer zone of pheromone traps along the periphery of the infested Northeast—if gypsy moth population densities within the zone are maintained below their "blow-out" level. In order to obtain population-density data and population-distribution patterns for the establishment of a barrier zone, disparlure-baited traps have been placed on a one-square-mile grid pattern in nonregulated areas of New York, Pennsylvania, West Virginia, and Virginia (Marx, 1972b).

Methods of control using disparlure concentrate on preventing male moths from finding female moths. However, these methods are suitable only for light infestations and involve mass trapping. There are two basic methods of control with disparlure. The first consists in luring the male to its destruction, and the second serves to confuse the male in his attempt to seek out a female.

Male moths may be lured to their deaths by the use of pheromone placed in traps with a sticky substance or an insecticide present. This method of control follows the general theory that if large numbers of traps are distributed and enough males are trapped, then a reduction in the population density will result. The number of moths caught in the field depends not only on the effectiveness of the lure, but also on the design and size of the traps and their placement. Granett (1973) has designed several traps with a greater capacity for surveying and trapping the number of moths in dense populations and for possible use in control programs. Several of his models of box-shaped traps are not dependent upon sticky surfaces for holding gypsy moths and have the potential of catching more than 1,000 moths each. Chemosterilants may be used along with disparlure in traps that allow access to the gypsy moth. Positively responding moths that come in contact with the chemosterilant or even feed on it would render themselves sterile and thus incapable of sustaining a growing population.

Disparlure may be used in the confusant or air-permeation method of control, but again only in light infestations (Beroza, 1960; Gaston et al., 1967; Shorey et al., 1967 and 1972; and Klun and Robinson, 1970). In 1963 R. H. Wright theorized that insects could be controlled by permeating the atmosphere with a sex-attractant chemical so that any additional quantity released by a female would be imperceptible. This would confuse the male or hinder his attempt to seek out a female by preventing normal responses and proper orientation to the female's pheromone source, thereby reducing the number of matings and the future size of the insect population. At high population densities, males can find females regardless of the amount of sex-attractant chemical used, owing to the great number of females present in the environment. This chemical, introduced as an external, artificial stimulus into the environment of the insect for the purpose of modifying or inhibiting the insect's proper behavior by eliciting or inhibiting an appropriate response, is termed *metarchon* (Wright, 1963). With disparlure as a model, if the threshold concentration for detection of the sex attractant by the male moth is taken as 1,000 molecules per mm^3 of attractant, then a concentration of 100,000 times higher than the threshold completely saturates the male antennae. Twenty kilograms or 100,000,000 molecules of disparlure would be required to saturate 600 km^3 of air.

Aerially applied microencapsulated formulations of disparlure that adhere throughout the forest canopy are capable of highly significant suppression of gypsy-moth mating in new infestations with low-population densities (Cameron, 1973; Cameron et al., 1974). It is important that the lure be distributed throughout the vertical profile of the forest instead of having it all on the forest floor. This distribution pattern can be achieved either with aerially applied granular corks impregnated with the synthetic pheromone or with a water slurry of microcapsules stabilized by a thickener, each containing a sticker so that material will adhere to the foliage. It has been suggested that 15 grams of disparlure per hectare may be sufficient to disrupt mating behavior in low-density populations to prevent population increases of the insect. The confusant formulations show promise for gypsy moth control in light infestations or low-population densities. Moreover, the approximate cost of disparlure confusant per acre for treating large areas of infestation is about one dollar (Beroza, 1973).

The confusant method and mass trapping also can be achieved by dropping traps baited with disparlure (Knipling and McGuire, 1966). However, the ratio of traps to females would have to be high to reduce the chances of males to locate females.

Disparlure also is expected to have little or no effect on reproduction of gypsy moth parasites in the field, Such has been demonstrated for the parasitic wasp, *Brachymeria intermedia* (Cameron and Rhoads, 1973). This is an important aspect to be considered if the synthetic attractant is to be used in an integrated biological control program. Both mass trapping and confusion methods of control using disparlure theoretically would become more efficient as the population declines. These depressed populations then could be managed by disparlure or other control agents.

At the present time disparlure provides an excellent method of survey and detection of the gypsy moth and a good method of combating low-density infestations. Disparlure may well become more effective as improvements in formulations and longevity are developed. Ultimately, disparlure may play a very significant role in an integrated gypsy moth management-and-control program if its use can be applied to high-density infestations. The synthetic sex attractant also may aid in spot-eradication programs and in the establishment of a pheromone-treated barrier for early detection and protection action against further spread of the insect. Disparlure also may help to reduce the area of general infestation; however, because the effectiveness of disparlure is reduced at high-population levels, it alone will probably not be effective where the gypsy moth is already causing great defoliation and economic damage. The chief advantage of disparlure may well be to provide time to explore other ecological avenues in dealing with high-density populations and the gypsy moth problem in general.

Table 13 - 16
EFFECTIVENESS OF DISPARLURE
—NUMBER OF GYPSY MOTHS CAPTURED—

ug Attractant per Trap	Approximate Age of Attractant (weeks)		
	1	6	12
	Natural extract*		
	11	6	0
	Disparlure**		
1.0	127	146	138
0.1	155	109	126
0.01	88	90	40

| 0.001 | 69 | 20 | 7 |

*Equivalent to ten tips per trap, the amount used in standard survey traps.

**Trioctanoin, 5 mg, added as keeper in each trap.

Disparlure is effective at very small concentrations and after exposure in the field for up to periods of 12 weeks.

(Courtesy: *Science;* Copyright 1972 by the American Association for the Advancement of Science, after Beroza and Knipling, 1972)

Table 13 - 15

MODEL OF THEORETICAL EFFECTS OF DISPARLURE-BAITED TRAPS

Day	Number of Traps	Unmated Females	Males	Ration of Traps to Unmated Females	Males Captured (No.)	Matings (No.)	Unmated Females Remaining	Daily Suppressed (%)
1	5000	100	100	50:1	98	2	98	98
2	5000	174	102	28.7:1	99	3	171	97
3	5000	228	102	21.9:1	98	4	224	96
4	5000	268	103	18.7:1	98	5	263	95
5	5000	297	104	16.8:1	98	6	291	94
6	5000	318	105	15.7:1	99	6	312	94
7	5000	334	105	15:1	98	7	327	93
8	5000	345	105	14.5:1	99	7	338	93
9	5000	354	105	14.1:1	99	7	347	93
10	5000	360	105	13.9:1	99	7	353	93

In theory disparlure-baited traps would intercept male moths before they can mate with female moths in an incipient population. The number of females mating in an uncontrolled population would be 1,000; the total number of females mating in the controlled population is 54. Mating is thus controlled by 94.6 percent.

(Courtesy: *Science*; Copyright 1972 by the American Association for the Advancement of Science, after Beroza and Knipling, 1972)

Table 13 - 17

RELATIVE ACTIVITY OF COMPOUNDS RELATED TO THE GYPSY MOTH ATTRACTANT

Compound (estimated to be 88% cis)	Molecular Weight	Structure	Activity *%
7, 8-Epoxy-2-methyloctadecane	282		100
3, 4-Epoxy-2-methyloctadecane	282		.1
3, 4-Epoxy-3-methyloctadecane	282		.001
6, 7-Epoxy-3, 4-dimethylheptadecane	282		.1
7, 8-Epoxy-3-methyloctadecane	282		5
7, 8-Epoxy-4-methyloctadecane	282		.5
8, 9-Epoxynonadecane	282		.02

From the table it can be seen that slight variations in molecular configuration result in great loss of biological activity of the chemical compounds as a gypsy moth attractant.

(*Source: Bierl and Beroza, 1970)

14

Physical Control Methods

There are several general hand methods of control that have been used against the gypsy moth by individual homeowners and government agencies from the early 1900s to the late 1930s. Applied separately or in combination, depending on forest growth, density of infestation, and other local conditions, these methods included the use of creosote, burlap and sticky bands, the destruction of egg masses, larvae, and pupae, and the cutting and burning of brush. Some of these methods, such as the use of burlap and sticky bands, capitalize on larval behavior to achieve control. Because of the considerable amount of hand labor needed and the extremely high costs of treatment, these methods have been outdated for use on large areas of woodlands; however, home and nursery owners are beginning to employ some of them in residential areas, small nurseries, and small, isolated woodland plots.

CREOSOTE

Painting gypsy moth egg masses with creosote kills the eggs without removing them from the trees or other objects on which they are laid. Creosote is a low-grade coal tar that has been impregnated with sufficient coal-tar pitch or lamp black to discolor and darken the egg masses. The discoloring of the egg masses serves to indicate those which have been treated. A good grade of creosote can be bought commercially in various quantities under various specification. Between 6.5% to 8.5% of coal-tar pitch by weight should be included in the creosote in order to produce a color suitable for staining gypsy moth egg masses (Burgess and Baker, 198). Regardless of the commercial preparation chosen, the mixture must remain fluid and workable at subzero temperatures. A pint of creosote can treat approximately 200 egg masses (Marx, 1970b).

Equipment needed for creosoting egg masses include brushes and poles, varying in height, to which the brushes can be attached to paint egg masses beyond normal reach, Hand-squeeze oil cans also can be used to apply creosote.

Creosoting is usually done between the first of August and the time of egg hatching in the spring. The treatment is most efficient when there is no snow on the ground or trees, and foliage is gone. It is under these conditions that the egg clusters are most readily visible for treatment.

Creosoting is effective on small shade trees and small woodland plots. However, there are two problems encountered when creosoting is relied upon for the destruction of egg masses. First, creosoting for effective control does not succeed if enough egg masses are left in the area unattended. Many egg clusters are laid in concealed or out-of-reach places under large rocks or upper tree branches where they are either unnoticed or inaccessible. These egg masses are usually left untreated and allowed to develop, resulting in damage to the site. Consequently, the effectiveness of creosoting is limited by the inability to find all egg masses. Second, the wind blows many young larvae from untreated areas to treated areas, which allows for the possibility of reinfestation.

BURLAP BANDS—LARVAL TRAPS

In certain infestations where spraying is impractable or objectionable, burlap bands or larval traps placed around the trunks of trees are often useful in controlling the gypsy moth. A large number of larvae can be trapped easily, removed, and killed with these traps. The use of larval traps in this manner capitalizes on the behavioral characteristics of the larvae, which usually seek shelter during hot, sunny days. When a band of burlap is attached to a tree, large numbers of larvae crawl beneath it to avoid the sunlight, heat, and their effects of dessication. From this band the larvae can be collected and crushed each day.

A burlap strip 8 to 12 inches wide is used to construct one form of a larval trap. It is placed loosely around the tree trunk approximately four feet above the ground, and a cord of twine is passed over the center of the burlap and tied to hold it in place. The upper half of the burlap is then folded down over the lower to make a double shelter.

A simple yet effective alternative form of larval trap can be constructed by folding a piece of heavy paper in half and tacking it to a tree.

STICKY-BANDS—LARVAL TRAPS

Sticky-banding as a method of control also utilizes the behavioral characteristics of larvae for their destruction. A sticky-band is made by placing a two- to four-inch-wide strip of cotton batting around a tree approximately four feet above the ground. Over the cotton a four- to six-inch strip of tar paper is wrapped firmly and tacked into place. Finally, a sticky material is smeared over the tar paper.

The sticky materials most often used may be obtained on the commercial market; they include tanglefoot, stiken, and "gypsy moth tree-banding material," TBM. These materials are composed of coal-tar oils and coal-tar pitch. Hydrated lime is contained in TBM. All are very effective in preventing larvae from crawling up the trunks of trees. TBM was developed by the Bureau of Chemistry, USDA in 1915 as a substitute for the German tree-banding material, *"raupentein,"* which was used experimently in the United States from 1892 to 1914 (Smulyan, 1923). TBM must be applied with a special gun or applicator in a thick, narrow band encircling the tree trunk, while tanglefoot and stiken can be applied by brush. All of these materials are expensive, and considerable labor is required to prepare and tend to the bands.

If these bands are placed on the trees in late April or early May before hatching of eggs and migrating of early instars, larvae are unable to cross the sticky-bands or crawl under the tar paper and cotton. Because the surface of the bands tends to dry and harden, it should be freshened with new applications of sticky material or by scratching the surface with a comb to bring up fresh sticky material. This should be done every seven to ten days, and bands should be inspected as often as possible from late May to mid-July.

Sticky-bands also may be applied in slightly different manners. The bark of the tree may be stripped, so that the banding material can be applied evenly in a thin layer with a brush. These bands prevent larvae from ascending the tree but leave temporary unsightly marks on the bark. Nurseries often apply a six-inch band of tanglefoot directly around tree trunks. This may not be so effective as the cotton-tar/paper-tanglefoot combination, and it is definitely messier. If proper treatment of egg clusters has reduced the number of larvae sufficiently, sticky-banding is a very effective measure when applied in small-scale programs.

Before 1896 some materials applied to tree trunks to prevent gypsy moth caterpillars from reaching tree foliage included gas tar, tree lime, and bird lime (Collins and Hood, 1920).

In exposed areas such as hill tops and mountain ridges, where there is a danger of infestation from windblown larvae and where egg masses have been found on the ground, sticky-bands formerly were placed on the trunks of trees to prevent early instars from climbing to the foliage to feed. Because of widespread infestations and the chance of wind-blown reinfestations, sticky-bands are no longer used in this manner. However, banding is effective on isolated trees or rows of trees where wind dispersal of larvae and their migration pose no serious threat.

Trapped larvae can be crushed or burned. If larvae are not directly killed, they are usually crowded in large numbers beneath the sticky-bands—a favorable condition for the development of "wilt" disease and starvation.

Although obsolete for large-scale operations, these methods of hand control, creosoting egg masses, and using burlap bands and sticky-bands, along with using lead arsenate and cutting and burning brush, were the chief control measures against the gypsy moth from the late 1890s to the development and use of DDT.

Today, the Northeast Forest Extermination Service at Hamden, Connecticut, has suggested the use of many of these hand-control methods for homeowner protection (Marx, 1974b). They suggest that individuals follow this procedure: (1) Scrape off and destroy all egg masses within reach of the ground; (2) have all egg masses sprayed from the ground with Bt; (3) remove all potential resting sites; (4) use burlap bands on all trees two inches or more in diameter; and (5) use disparlure traps to capture male adults.

15

Quarantine

In order to protect the economy of the United States, a quarantine regulating shipments of various materials is maintained against the gypsy moth in an effort to prevent the further spread and establishment of this pest in uninfested areas. Materials regulated are considered as potential carriers of the egg, larval, and pupal stages of the gypsy moth. Quarantine usually is enacted by federal and state agencies to prevent a limited infestation from spreading throughout the ecological limits of the species.

The first important regulatory legislation in the United States directed against insect pests was passed in 1877, when four states enacted legislation on pest control. The first federal law concerned with insect pests was passed by Congress in 1884. With the passage of the Federal Insect Pest Act of 1905, the Plant Quarantine Act of 1912, and their later revisions, the federal government was able to regulate interstate movement of articles that might spread insect pests. In addition, the Plant Quarantine Act was authorized by the Secretary of Agriculture to enforce necessary regulations to protect the agricultural economy of the nation. It is through the Plant Quarantine Act that current programs of handling insect pest problems have developed.

The general authority and responsibility of the USDA to

enforce quarantine regulations and to develop and execute programs to eradicate, control, and suppress insect pests were further strengthened by the Organic Act of 1944 and the Federal Plant Pest Act of 1957 (Koski, 1971b). Under these laws the Agricutural Research Service, Plant Protection and Quarantine Division of the USDA is responsible for enforcing the provisions of the Domestic Plant Quarantine Act No. 45, Gypsy Moth Quarantine (Warner, 1973). These acts also permit cooperation among federal and state agencies in combating the gypsy moth and provide for emergency action and issuance of regulations. The need for greater federal involvement to protect forest-related resources resulted in the Forest Pest Control Act of 1947, which gave the United States Forest Service responsibility for gypsy moth control.

Effective quarantine measures require the cooperation of state and federal agencies. State agencies responsible for enforcing quarantine regulations to prevent the spread of the gypsy moth within state boundaries have had additional legislation passed by their respective assemblies to supplement federal law and guidance. In 1933 Pennsylvania passed the Gypsy Moth Quarantine Act. This law made it necessary for susceptible material in quarantine areas to be inspected and certified as free from the gypsy moth before it could be moved (Nichols, 1962). This act was later supplemented by the Pennsylvania Plant Pest Act of 1937. Cooperation from commercial carriers and prospective shippers in complying with quarantine measures also has been sought, along with public information programs.

Under present federal gypsy moth quarantine regulations, no timber or timber products, nursery and ornamental stock, stone and quarry products, junk, mobile homes, trailers, or campers may be moved from points within the quarantine area to points outside without first being inspected by federal or state agricultural agents. In addition, regulated articles in transit must have a federal or state certificate of inspection. Regulated articles may be moved from infested areas if the following qualifications are met: (1) Approved treatment is applied to infested materials; (2) articles have been examined and found to be free of the gypsy moth; or (3) the articles have been grown or handled in such a manner that no infestation can be transmitted.

In addition to regulation control, present quarantie laws also enforce the application of pest-control measures when properties are endangered by the inaction or carelessness of others, combat the gypsy moth on public and private lands, and prevent misbranding of insecticides used against the insect.

Quarantine is imposed on those areas known to be infested and on areas that have been exposed to infestations. At the present time 12 states are under gypsy moth quarantine, 7 of which are regulated in their entirety. These states are Connecticut, Delaware, Maine, Maryland, Massachusetts, Michigan, New Hampshire, New Jersey, New York, Pennsylvania, Rhode Island, and Vermont. Extensive federal-state quarantine operations and control programs have somewhat confined the gypsy moth to the Northeast, but in recent years the insect has increased to high-population levels and is spreading from the Northeast into many parts of the southern and midwestern woodlands.

Quarantine action against the gypsy moth is approached on a containment basis, for there is no attempt made to eradicate the insect throughout the infested areas of North America. Usually control is applied in the form of suppressive pesticidal sprays along selected portions of the periphery of infested areas in the Northeast to retard the spread of the gypsy moth, while quarantine action is maintained over those regulated articles to prevent long-distance dispersal of the insect. Constant surveillance is necessary to determine the dispersal of the insect and to update adjustments in the dimensions of the quarantine areas.

Currently, the long-distance dispersal of the gypsy moth is being aided greatly by the use of mobile homes and camping trailers. These vehicles are *the* most dangerous regulatory problem and present the greatest threat to gypsy moth containment efforts due to their large numbers and flexibility of movement. Their ease of movement makes it almost impossible for the USDA to provide adequate inspection service to all mobile homes and camper trailers in regulated areas. Mobile homes began to receive national recognition as a regulatory hazard in the late 1960s and early 1970s. Vehicles present in infested areas during June and July carry egg masses, larvae, and pupae to distant, noninfested areas. La Fleur (1973) has cited an increase in the popularity of mobile homes, the construction of hundreds of mobile home parks, and increased usage of mobile homes by military personnel as factors causing long-distance dispersal of the gypsy moth.

Recreational vehicles and campers have a similar effect upon long-distance dispersal of the insect. Trap catches have been made in close association with recreational areas in Alabama, Washington, D.C., Michigan, North Carolina, South Carolina, Tennessee, Virginia, West Virginia, and Wisconsin (La Fleur, 1973). This undoubtedly indicates that mobile homes and recreational vehicles represent a serious threat to gypsy moth control measures.

REGULATED ARTICLES
THAT MUST BE MOVED UNDER CERTIFICATE

1. Trees, shrubs with persistent woody stems, and parts thereof, EXCEPT seeds, fruits, and cones.

   Trees and shrubs, and parts thereof are exempt if grown in a greenhouse throughout the year and so labeled on shipping container.*

   Cuttings and scions with stems no greater than one-half inch in diameter are exempt.*

Parts of trees and shrubs that have been dried, pressed, waxed, lacquered, varnished, or similarly surface-treated are exempt.*

Christmas trees are exempt.*

Boughs and Christmas greenery are exempt.*

2. Timber and timber products, including but not limited to lumber, planks, poles, logs, cordwood, and pulpwood.

Lumber is exempt if dressed or sawed four sides with ends clipped and free of surface bark, or if kiln dried, provided such lumber is shipped directly after processing and the waybill or other shipping document is marked to show that the lumber was shipped immediately after processing.*

Manufactured wood products, such as shingles, flooring, furniture, handles, etc., are exempt.*

3. Stone and quarry products.

Stone and quarry products are exempt if processed by grinding or pulverizing.*

4. Mobile homes, recreational vehicles, and associated equipment moving from hazardous parks or recreational sites.

*Exempt if not exposed to infestation after the prescribed handling.

(Courtesy: USDA, 1972)

16

Silvicultural Control

Silvicultural control of forest insects consists of the regulation of forest growth to promote unfavorable food conditions for pest insects. This method of developing insect-resistant forests is sometimes the most practical type of long-range forest insect control if insect infestations are not widespread. Control is achieved by manipulating the quantity and quality of insect food through the regulation of forest composition, density, vigor, and age of the stands. In essence, the quantity of favored or susceptible host plants is reduced, while the unfavored or resistant plants are encouraged through a program of selective thinning. The vulnerability or susceptibility of trees to damage decreases with the growth rate of the tree, that is, trees growing most rapidly have the greatest capacity to recover (Balch, 1958). However, it is best to prevent the establishment of pure stands, since they often may be more susceptible to insect damage than mixed stands. In pure or nearly pure stands of favored hosts, clear cutting and planting of unfavored hosts is the best method of building resistant stands if operational expenses can be met.

Silvicultural regulation often varies greatly from one area to another. The existing condition of each stand and combination of species must be considered along with such controlling factors as site, soil, market, timber value, and degree of infestation or possible infestation.

The use of woodland management to combat the gypsy moth originally was proposed by Clement and Munro (1917) and renamed silvicultural control by Behre, Cline, and Baker (1936). In consideration of the silvicultural control method, the classification of plants into four groups previously mentioned was used as a guide for thinning and eliminating the most favored hosts and retaining and encouraging the less vulnerable and resistant hosts.

Mixtures of species from Group 1 and Group 2 are susceptible only when the proportion of foliage in Group 1 is high enough to allow a large number of larvae to enter late instars. In such mixtures trees in Group 1 are usually defoliated completely before those in Group 2 are attacked severely.

In mixtures of Groups 1 and 3 the defoliation of trees in Group 3 is also dependent upon the proportion of Group 1 trees present. If sufficient foliage from Group 1 is present to allow larvae to reach the third instar, trees in Group 3 may be severely attacked. Woodland stands composed predominantly of trees from Groups 2 and 3 are fairly resistant but can be made more resistant to gypsy moth attack. This can be done if favored threes in Group 1 growing in the stand are removed, since they are essential for young larvae to survive.

Mixtures of Groups 2, 3, and 4 are seldom defoliated. Stands composed chiefly of Group 4 trees are in no danger of serious damage, regardless of the presence of some Group 1 trees.

The susceptibility of a woodland stand to gypsy moth infestation and damage depends not only on the abundance of favored food hosts, but also upon the site and stand condition. On dry sites the proportion of oaks, gray birch, and aspen were usually reduced, and the growth of hemlock, white pine, red maple, ash, black locust, and hickory encouraged (Bess et al., 1947). On moist sites the proportion of favored species was not reduced so much as on dry sites. Bess, Spurr, and Littlefield (1947) indicated that such dry sites as sand plains and rocky ridges—those which support open stands—are the most susceptible areas, especially where land abuse is common.

Sites having an adequate supply of moisture and organic matter are more resistant to attack and are characterized by relatively vigorous growth. These moist sites favor the presence of vertebrate predators such as the deer mouse, the white-footed mouse, and the short-tail shrew, all of which contribute to the relative resistance of the forest stand by feeding on gypsy moth larvae. On sites where most of the forest growth is of low commercial value, some immediate relief can be obtained by the adoption of a thinning program to remove the most favored trees. This thinning, together with the encouragement of the growth of nonfavored hosts, can help in preventing damage and spread of the insect. Gradual removal of favored hosts, especially the oaks, through logging operations or improvement cuttings, also reduces the damage caused by the gypsy moth.

Unfortunately, if silvicultural control of the gypsy moth is to be used not only to curtail infestations in infested areas but also to prevent the spread of the insect, control must be applied on a very intense and costly scale because of the present forest conditions in many infested areas.

The present gypsy moth-infested forest regions,[2] which cover several million acres, abound in Group 1 trees in solid stands and in more or less diluted mixtures. This condition is not good from any forest-management viewpoint, especially when silvicultural control is considered. Because of the widespread distribution of infestations and woodland compositions in the Northeast and in the susceptible woodland areas of the South, silvicultural control of the gypsy

moth is no longer considered an effective control measure
on either large- or small-scale operations. Once of great
importance in forest management, the control technique
was abandoned in the late 1930s. In addition to the widespread distribution and susceptibility to attack, each stand
and combination of tree species presents its own problems,
further complicated by financial and operational costs.

Many of the original forest types in the northeastern
United States contained smaller percentages of favored
gypsy moth hosts than the forest types of today. However,
for the past 80 or 90 years, excessive cutting and logging
operations in the original forests, destructive forest fires,
and the disappearance of the American chestnut, *Castanea
dentate* (caused by the chestnut blight fungus, *Endothia
parasitica*), have produced the unnatural forest types that
exist today in the Northeast. These unnatural types were
the major stands benefiting from dramatic environmental
changes. Today, in Pennsylvania alone, because of these
alterations in the forest environment, over ten million acres,
or 60% of the forest area in the state, contain practically
solid oak stands.

17

Survey

The detection, occurrence, and estimation of numbers of the gypsy moth are necessary for the success of quarantine and integrated control measures against the insect. There are two basic means of obtaining information about the presence and abundance of the gypsy moth: the manual type of scouting survey and the trapping or detection survey.

An additional survey method, aerial observation, can be used to spot defoliated areas quickly. Aerial surveys provide information on relatively large areas of woodlands and the extent and intensity of gypsy moth defoliation; they also save time and money when defoliation is detected early. Aerial surveys along with ground surveillance programs are mapped on all forest lands to detect pest infestations and trends.

Before 1940 surveying by scouting was the chief means of obtaining information about gypsy moth infestations. Labor was abundant and inexpensive, and scouting and creosoting of egg masses were accomplished for approximately twenty cents per acre.

By 1941 with U.S. entry into World War II, scarcity of labor reduced budgets caused a gradual transition in survey programs to those that exist today.

The cost of scouting has risen to well over one dollar

per acre and is done only on a selective basis. Following positive trap catches of male moths, scouts may patrol an area to determine the site and relative size of an infestation. Scouting also may be done to analyze spraying programs, to supplement trapping work, and to inspect critical areas such as camp sites or mobile home courts, where an infestation might easily become established (Nichols, 1962).

Scouting is best conducted from late fall to mid-spring, when tree foliage is absent and egg masses, pupal cases, and cast skins of larvae can be seen easily. During scouting surveys special attention is given to such objects as tree trunks and large tree limbs, loose bark, rocks, and rubbish. Host trees preferred by the larvae receive additional attention.

Nichols (1962) has described three types of scouting surveys : spot, strip, and intensive.

The spot survey is used when few men are available and the area to be surveyed is relatively large. Selected localities and specific trees are examined in the suspected infestation.

The strip survey is the most commonly practiced method when a large infestation is suspected and a large number of men are available. Men are spaced from 10 to 100 feet apart in a straight line and marched through a suspected infestation. The spacing varies with the terrain, and trees to be examined are chosen by their susceptibility to the gypsy moth. Moving men back and forth through the woodlands allows extensive areas to be covered in one day.

A modified method of strip surveying is roadside scouting. Commonly used, it consists of examining trees in the open country, around houses, and in wooded areas in strips of 50 to 100 feet wide along roadways. If light infestations are suspected, only particular trees such as oaks are exaxmined. However, if heavy infestations are possible, as many trees as possible are examined for gypsy moth evidence.

When a small infestation is suspected, the intensive survey method is used. Scouting consists of tree-to-tree examination along with examination of rocks, junk, and leaf litter.

TRAPPING

Sex-attractant traps are used to locate areas infested with the gypsy moth, to monitor the effectiveness of spraying programs, and to determine relative numbers in gypsy moth populations. Trapping surveys for gypsy moth detection were first conducted in 1934. With increased improvements in attractants and traps, trapping procedures gradually replaced scouting and have since proved to be effective in obtaining gypsy moth information for control.

The first survey trap consisted of tanglefoot smeared on the side of a tree with the natural sex attractant attached in a tin can, placed in the center of the tanglefoot (Nichols, 1962). The model that followed was made of a metal cylinder with the attractant suspended by fine wire to the inside of the cylinder. Lining the inside of the cylinder was a sheet of removable paraffin or wax paper smeared with tanglefoot. The ends of the trap were covered with inverted waxed-paper cones having holes large enough to allow the male moth to enter the trap.

In 1960 a green, disposable, eight-ounce paper trap was developed, similar to a paper cup in construction and coated inside with polyethylene (Nichols, 1962). The trap was manufactured with a handle for attachment to a tree and had entrance holes in the lid and bottom. These openings also ensured air circulation and water drainage. The traps cost approximately two cents each, and they also eliminated the yearly trap-cleaning expenses required by the metal traps and were easier to assemble, maintain, and store. Tanglefoot again was applied to the inside surface of the traps, where its absorption into the cup was prevented by the polyethylene coating. A cotton dental pad, 12 mm in diameter by 6 mm in thickness, containing the sex attractant was placed on a small square of gummed acetate and stuck into the tanglefoot at the top center of the trap. Gyplure was used first as the sex attractant in these traps and was replaced later by disparlure.

Starting in 1969 new experimental forms of larval traps were used in selected areas of most states in the Northeast.

In the early 1970s gypsy moth traps consisted of yellow, weather-resistant, cylindrical cardboard, 5.3 cm in diameter and 10 cm long, with clear plastic ends having 2.5 cm openings for male moths to enter. These traps were used along with the green disposable traps and eventually replaced them. The cylindrical traps also were lined with tanglefoot and contained the synthetic sex lure. Disparlure was mounted into the trap on a cotton wick stuck to the top center of the trap.

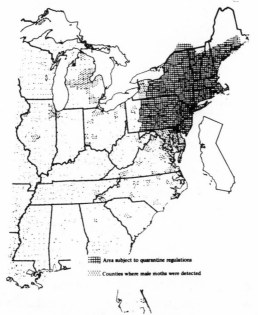

Gypsy Moth Survey, 1973. *Courtesy: USDA, 1973c*

Gypsy moth cylindrical trap with positive catches. Traps are baited with disparlure and provide data on the insect's spread and population densities. *Courtesy: USDA.*

Beginning in 1975 red, triangular Delta traps were used by the USDA and Commonwealth Agriculture Departments. The trap is precoated with tack trap and measures 10 cm on each side with a 3-cm, triangular opening at each end. Disparlure is suspended from the inside roof of the trap in a Herculon wick, which consists of a polymer sandwich of several layers of plastic material (Marx, 1975). The Herculon wick outperforms the dental pad and cotton wick (Marx, 1974a).

As little as 0.005 ug of disparlure has been used to bait some survey traps (Beroza et al., 1971b). Males responding to the attractant enter the trap and get stuck to the sticky coating material. Captured moths can be removed from the adhesive by soaking them in methylene chloride for 30 to 60 minutes. Present survey traps are baited with approximately 100 micrograms of disparlure each. This amount is over 50,000 times that of the natural lure formerly used to bait survey traps.

Agricultural scientists are currently designing a new survey trap containing a chemical killing agent that would eliminate the need for tanglefoot.

Disparlure-baited traps are usually placed in the field between mid-June and July 1 and remain there until September 1. It is extremely important that all traps be placed in the field before male moths emerge. Traps are hung four to five feet from the ground, either tacked directly to the tree trunk or suspended by wire from tree limbs. If traps are placed near roadside or woodland edges, they are usually set 10 to 100 feet back from such edges.

A survey trapper can handle from 30 to 50 traps a day, depending on the type of trapping grid to be laid and on the nature of terrain and roads to be traveled. Each trap is inspected every three to four weeks. Inspection consists of collecting any possible "positive" catches for further identification, removal of other insects, and reworking the sticky inner coating.

There are two basic types of trapping surveys: grid trapping and roadside trapping.

Grid trapping is performed by two individuals and is used in areas suspected of infestation or where infestation could result from nearby populations. This survey is done by placing traps at definite periodic intervals on a grid layout.

Roadside trapping is usually accomplished by one individual who places traps at one-mile intervals or less along all roadsides and trails in a suspected area. Intensive trapping requires that traps be placed at ⅞-mile intervals in lines one mile apart. If very large areas are to be trapped at densities ranging from one per nine square miles to one per fifty square miles. Roadside trappings are conducted as a general method of survey where knowledge of the presence and spread of the gypsy moth is required for a large and generally uninfested area. Intensive trapping surveys are being conducted in many states bordering the generally infested Northeast, throughout Nova Scotia and New Brunswick, and throughout camping areas west to the Rockies.

The capture of a single male moth does not necessarily indicate the presence of an infestation in the immediate vicinity of the trap site; for the adult may be carried by the wind 20 to 30 miles away from an infestation and is capable of traveling 100 miles. Isolated positive catches within a 100-mile radius of an infestation do not represent a new infestation, but repeated yearly captures or the presence of multiple catches in a single trap usually indicates an infestation within a one- to two-mile radius of the trap site.

In an effort to prevent small infestations from flourishing, large-scale trapping programs are conducted so that infestations can be detected early. It appears that these programs will be continued as long as the gypsy moth remains a threat to major woodlands.

In 1972 an intensified trapping program was conducted to obtain information concerning the relative abundance of dispersing gypsy moth populations in newly infested areas along the periphery of the generally infested areas of the Northeast and possible remote infestations. Approximately 120,000 traps were used in this program in 38 states and the District of Columbia. Male moths were trapped in 118 counties in 12 states and the District of Columbia—the first time that moths were trapped in 55 of these counties (Katsanos, 1973). In 1972 positive catches were made for the first time in Iowa, Tennessee, and West Virginia; in 1975 another intense trapping survey was conducted in Illinois, Indiana, Michigan, Ohio, Virginia, West Virginia, Delaware, Maryland, Minnesota, and Wisconsin.

FORECASTING DEFOLIATION

Today there are several reasonably acurate methods of estimating gypsy moth egg-mass densities (Patil and Stitler, 1973). These estimates can be used to forecast defoliation based on the scouting of all viable egg masses visible in randomly chosen localities. Egg-mass density is the most important factor in evaluating the possible damage from gypsy moth attack in a specific area. The number of viable egg masses, their relative size, the proportion of trees present that are favored hosts, and climatics conditions are considered in forecasting the expected degree of defoliation. However, there is a need for a standard method of estimating egg mass density and degree of expected defoliation.

Forecasting defoliation can be done at any time between May and August but is usually determined after the leaves have fallen, when egg masses are seen easily. In heavily defoliated woodlands it is difficult to estimate egg-mass density, for older egg masses remain on the bark of trees for several years. Although older egg masses are darker in color, it is difficult to single out these masses when large areas are to be covered in a limited period of time.

As a general rule, an average of 500 egg masses of normal size per acre in woodlands composed of at least 50% favored hosts usually results in heavy defoliation, while 1,000 egg masses per acre in the same woodlands results in complete defoliation (Marx, 1970a). Some egg mass densities in northwestern New Jersey and eastern Long Island have reached 20,000 per acre. Currently, information concerning number of egg masses, stand composition, and climatic conditions is being fed into computer models to obtain estimates of possible gypsy moth damage and control procedures.

18

Management

The primary objectives of present forest-management programs are to provide tree-foliage protection from gypsy moth defoliation and to give relief to residents in infested areas from the nuisance problems posed by the invasion of migrating larvae (Nicholas, 1973). Federal, state, county, township, and municipal governments often must combine efforts to reduce damages caused by the gypsy moth.

According to Cobb (1968), there are two components in an evaluation of a gypsy moth forest problem. First, a biological analysis is necessary to determine the possible damage that may result from gypsy moth attack unless control measures are taken, and to determine the control methods to be taken. Second, an economic analysis is performed to assess the value of the resources at stake and the cost of possible control measures. These two analyses are considered to determine if pest-control management is required.

Biological analysis of the infested area includes information on the biology and ecology of the gypsy moth; its population density and possible size fluctuations; the size, density, and variety of host plants on the site; and projected damage. Gypsy moth population structuring in the infestation is the critical factor considered in determining the need for control. Available control measures are reviewed with respect to the safety, efficiency, and cost of each measure. If the biological analysis indicates possible damage to be minimal, the situation is kept under surveillance; if, however, the analysis indicates the damage to be serious, an economic evaluation is conducted.

The economic analysis considers the impact of the gypsy moth upon timber value, wildlife, recreation, water, scenery, and the woodland environment in general. If expected damage far exceeds the cost of suppression or control benefits, a forest pest control program is initiated. A general scheme of the gypsy moth's economic impact can be evaluated in the following manner: The value of gypsy moth control is equal to the losses incurred during unchecked periods of infestation plus environmental risks, minus control costs and environmental control risks produced by the control measure (White, 1973).

Regardless of the method used, the efficacy of control against the gypsy moth ultimately depends upon the quality of information that was analyzed for selection of the control measure. To this end, scientists are working toward the development of a two-stage growth simulator (Valentine, 1973). The first stage of the simulator is a mimicry of the population dynamics of the forest stand containing host species vulnerable to gypsy moth attack. Input materials consist of tree species and composition, stand structure, climatic conditions, productive potential of the soil, and natality and mortality processes of the stand. This input structuring is based on the assumption that changes in the forest environment can be modeled as a function of its present state and random components (Valentine, 1973). The second stage of the simulator monitors the effects of defoliation in the stand. Such factors as increased susceptibility to pathogens, exhaustion of food storage, reduced productivity, and mortality are included.

The prototype simulator developed for population dynamics of tree species in the northern hardwood forest is being modified to apply to different forest types and changing forest conditions.

Most large control programs involve state and federal cooperation. These programs are reviewed by state and federal health officials, wildlife biologists, and agencies that may be affected by such pest control operations. Large-scale proposals involving federal assistance are screened by the USDA, Agricultural Research Service, and the USFS. If proposals are agreed to and funds available, the programs are presented to the Federal Committee on Pest Control, consisting of representatives from the Departments of Agriculture, Defense, Interior, and Health, Education and Welfare (Cobb, 1968).

Both the USDA and the USFS participate in cooperative gypsy moth control activities involving the prevention of gypsy moth dispersal and the suppression and eradication of new infestations. These two federal agencies are responsible for regulatory activities, detection surveys, control and eradication programs in new infestations, methods development and investigation, and suppressive efforts for the protection of forest resources. Little federal involvement is demonstrated in the management of gypsy moth populations in areas of general infestation where the insect is established, for these areas are considered state management problems.

In order to meet the demands of control against the gypsy moth, the Pennsylvania Department of Environmental Resources has established an order of priorities as a guideline for treatment of infested areas on large-scale spraying programs (Nichols, 1973). Top priority consists of treating forested parks, public recreational areas, special use areas, historic and natural sites, and nonprofit campgrounds. Second in priority are forested communities or rural residential areas. The last priority includes privately owned recreational or high-value forested areas.

The majority of one-component control systems, such as chemical spraying measures, lack the flexibility necessary for continued suppression and regulation of the gypsy moth. One-component systems have never worked in suppressing the spread of the gypsy moth in the generally infested Northeast.

Although the gypsy moth cannot be eradicated with present options, most agencies believe that the insect can be controlled and regulated as a native pest by use of integrated control programs. In order to eradicate an insect, it must have a slow means of dispersal, a limited area of distribution, and at least one stage of its life cycle susceptible to an eradication method (Nichols, 1962). Although the gypsy moth has a relatively slow rate of dispersal, it is widespread throughout 100 million acres of woodlands. Its dispersal also is being aided by the transportation of its egg masses, larvae, and pupae on mobile vehicles and commercial materials. Even with all four stages of the insect vulnerable to some form of treatment, the combination of technical control problems, rising costs, and the insect's populaion dynamics and biology make all attempts to eradicate the gypsy moth virtually impossible.

For gypsy moth control and regulation to succeed, simple empirical measures of management have been expanded to a system of integrated control based on principles of applied ecology. This system is a combination of chemical techniques with one or more other control procedures, resulting in a single, unified pest-control program offered with respect control problems, rising costs, and the insect's population dynamics of the gypsy moth. Many believe that diversified program of integrated control may reduce the use of insecticides by 40% (Sanders, 1975).

The need for integrated control led to the formation of

the National Gypsy Moth Advisory Council in 1969. This Council consists of representatives of state and feredal agencies, universities, industries, and conservation groups. Other organizations such as the Gypsy Moth Research, Development and Application Program and the National Three-Bug Program have been formed and engaged in gypsy moth control work and research.

Currently, there are four types of federal-state cooperative programs that involve regulation and quarantine, control, methods improvement investigation, and survey. Present federal-state goals are to retard the spread of the insect, to localize infestations outside the generally infested Northeast, and to treat and prevent serious gypsy moth damage.

The New Jersey Department of Agricuture is employing a pest-management program of integrated control against the gypsy moth. The objective of this program is to prevent tree mortality in residential and heavily used recreational areas through both chemical and biological means (Fringer, 1973). This program provides for quick, short-term foliage protection by short, residual insecticide usage until effective biological control can be achieved. However, many acres of trees are expected to be lost before the collapse of the gypsy moth occurs. This type of integrated control program is gradually replacing many old control programs on both state and federal levels.

Integrated control is expected to have far-reaching demands, illustrated by the use of NASA spacecraft in gypsy moth control programs. The Advanced Technology Application Corporation has submitted a request to NASA for satellite surveillance to photograph and record gypsy moth defoliation by the Earth Resources Technology Satellite.

Hopefully, the main thrust of present control programs can buy time for research and methods development to pioneer new technologies that will control the damage and spread of the gypsy moth to a more significant degree.

19

Future

The future is not bright for those invoved with and concerned about the gypsy moth, for it appears that the spread of the insect and its resuting damage can not be arrested by present means of control. Investigations will continue for better survey- and population-estimating methods, and studies will be made of the insect's impact on the forest environment, larval and adult behavior patterns, mortality-causing agents, and successful use of genetic or pheromonal control against the gypsy moth. What may come of these studies is uncertain, but one factor that will remain constant for both the investigations and control of the gypsy moth is their continued increase in cost with little relief from the insect's damages as it spreads west and south throughout the eastern United States.

Control programs that have failed in the past will continue to do so if used against the gypsy moth in large-scale operations. Physical, silvicultural, and chemical control programs have failed because of the costs and logistical operational problems presented by the large areas of gypsy moth infestations and the insect's population dynamics.

Biological control, both parasitic and microbial, and sex pheromone control are only beginning to show positive signs of restraining the gypsy moth population, but their effectiveness in large-scale operations is questionable. Furthermore, factors such as repeated applications and high costs that have limited the use of chemical sprays also may limit the use of microbial sprays. Unless the sex phermone can be utilized in large areas of infestations, it may serve only as a means for securing more time during which additional control measures may be developed.

New, integrated control programs involving two or more methods of gypsy moth control are being prepared and released against the insect, but they too must prove to be significantly effective in large infestations. Information concerning their success or failure is just beginning to be received and analyzed.

In short, 100 years after the gypsy moth's introduction into North America and continuous control and eradication measures, the gypsy moth has not been eradicated or controlled, nor has its impact been reduced to that of a native woodland pest.

Unless effective, large-scale control measures against the gypsy moth are developed, adjustments will have to be made to compensate for its impact. Unfortunately, it may be that the biology and population dynamics of the gypsy moth are too well evolved for present technology to control.

GYPSY MOTH QUARANTINES

U. S. DEPARTMENT OF AGRICULTURE
ANIMAL AND PLANT HEALTH INSPECTION SERVICE
PLANT PROTECTION AND QUARANTINE PROGRAMS
AND CANADA DEPARTMENT OF AGRICULTURE
COOPERATING WITH AFFECTED STATES.

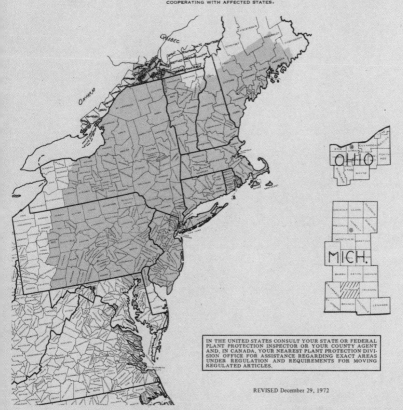

IN THE UNITED STATES CONSULT YOUR STATE OR FEDERAL PLANT PROTECTION INSPECTOR OR YOUR COUNTY AGENT AND, IN CANADA, YOUR NEAREST PLANT PROTECTION DIVISION OFFICE FOR ASSISTANCE REGARDING EXACT AREAS UNDER REGULATION AND REQUIREMENTS FOR MOVING REGULATED ARTICLES.

REVISED December 29, 1972

COUNTIES ENTIRELY COLORED ARE COMPLETELY REGULATED;
COUNTIES PARTIALLY COLORED ARE PARTIALLY REGULATED

GENERALLY INFESTED AREA–STATE, FEDERAL, AND CANADIAN REGULATIONS
(Eradication Treatments Not in Progress or Planned)

SUPPRESSIVE AREA–STATE, FEDERAL, AND CANADIAN REGULATIONS
(Suppressive Treatments in Progress or Planned)

STATE REGULATIONS ONLY
(Suppressive Treatments Planned Where Necessary)

ERADICATED–REGULATION REMOVED

RESTRICTIONS ARE IMPOSED ON MOVEMENT OF REGULATED ARTICLES FROM A REGULATED AREA AS FOLLOWS:
1. FROM RED INTO OR THROUGH GREEN OR WHITE.
2. FROM GREEN INTO OR THROUGH WHITE.
3. GREEN INTO GREEN.
4. WITHIN GREEN.*
5. FROM BLUE INTO ANY OTHER AREA.**

*IF REQUIRED BY AN AUTHORIZED INSPECTOR.
**IF REQUIRED BY APPROPRIATE STATE QUARANTINE OR BY AN AUTHORIZED INSPECTOR.

Courtesy: USDA.

Bibliography of Works Cited

Abbott Laboratories. 1971. *Dipel, a Biological Insecticide.* North Chicago, Illinois. 24 pp.

Acree, F., Jr. 1953. Isolation of gyptol, the sex attractant of the female gypsy moth. *J. Econ. Entomol.* 46:313.

————. 1954. Chromatography of gyptol and gyptol ester. *J. Econ. Entomol.* 47:321.

Agricultural Research. 1973. *The Looming Biological Battle.* September issue. 3–6.

Amirkhanova, S. N. 1962. Nutrient substances in the leaves of healthy and weakened food plants for the gypsy moth. (Russian) *Issled. ochagov. vredit. lesa. Bashkirii UF A* 2:81–95. (in Russian). (Biol. Abstr. 45:75486).

Anonymous. 1953. *Distribution Maps of Insect Pests:* Ser. A, Map no. 26, *Lymantria dispar* (L.). Comm. Inst. Ento., London.

App, A. A., and Granados, R. R. 1972. Effect of nuclear polyhedrosis virus infection on polyribosome content of gypsy moth larvae (*Porthetria dispar*). *Exptl. Cell. Res.* 519–24.

Atkins, E. L., Jr.; Anderson, L. D.; Nakakihara, H.; and Gregwood, E. A.; 1970. Toxicity of pesticides and other agricultural chemicals to honey bees. *Cal. Agr. Ext. Sta. Bull.* M–16. 25 pp.

Balch, R. E. 1958. Control of forest insects. *Ann. Rev. Entomol.* 3:449–68.

Behre, C. E.; Cline, A. C.; and Baker, W. L. 1936. Silvicultural control of the gypsy moth. *Mass. Forest and Park Assoc. Bull.* 157. 16 pp.

Beroza, M. 1960. Insect attractants are taking hold. *Agric. Chem.* 15:37–40.

————. 1957. Nonpersistent inhibitor of the gypsy moth sex attractant in extracts of the insect. *J. Econ. Entomol.* 60:875–76.

————. 1971. Insect sex attractant. *Am. Sci.* 59:320–25.

————. 1973. Gypsy moth sex lure (disparlure) for control. Speech given at Gypsy Moth Technical Work Conference. Beltsville, Maryland. February 6–7.

————; Bierl, B. A.; Knipling, E. F.; and Tardif, J. G. 1971a. The activity of the gypsy moth sex attractant, disparlure, vs. that of the live female moth. *J. Econ. Entomol.* 64:1527–29.

————; Bierl, B. A.; Tardif, J. G.; Cook, D. A.; and Paszek, E. C. 1971b. Activity and persistence of synthetic and natural sex attractants of the gypsy moth in laboratory and field tests. *J. Econ. Entomol.* 64:1499–1508.

————, and Knipling, E. F. 1972. Gypsy moth control with the sex attractant pheromone. *Sci.* 177:19–27.

Bess, H. A. 1961. Population ecology of the gypsy moth, *Porthetria dispar* L. (Lepidoptera: Lymantriidae). *Conn. Agric. Exp. Sta. Bull.* 646. 43 pp.

————; Spurr, S. H.; and Littlefield, E. W. 1947. Forest site conditions and the gypsy moth. *Harvard Forest Bull.* 22. 56 pp.

Bierl, B. A., and Beroza, M. 1970. Potent sex attractant of the gypsy moth: its isolation, identification, and synthesis. *Sci.* 170:87–89.

Bitzer, J. H. 1971. Some observations of tree mortality caused by gypsy moth defoliation on Chestnut Ridge, Monroe County, Pennsylvania. Misc. paper. Penn. Dept. Envirn. Res., Bur. Forestry. 3 pp.

Block, B. C. 1961. Behavioral studies of the responses of gypsy moth males (*Porthetria disparia* L.) to the female sex attractant and related compounds. Presentation to Eastern Branch Ent. Soc. Am. Meeting. N.Y., N.Y.

Brown, A. W. A. 1967. Insecticide resistance comes of age. Presidential address delivered at the opening session of the annual meeting of the Entomological Society of America. N.Y., N.Y. November 27.

Brown, G. 1968. The gypsy moth, *Porthetria dispar* L., a threat to Ontario horticulture and forestry. *Proc. Ent. Soc. Ont.* 98:12–15.

―――. 1973. Gypsy moth in Canada—report for 1972. Speech given at Gypsy Moth Technical Work Conference. Beltsville, Maryland. February 6–7.

Burgess, A. F., and Baker, W. L. 1938. The gypsy and browntail moths and their contral. USDA Circular no. 464. 38 pp.

Cameron, E. A. 1973. Disparlure: a potential tool for gypsy moth population manipulation. *Bull. Ent. Soc. Am.* 19(1): 15–19.

―――, and Rhoads, L. D. 1973. Oviposition by *Brachymeria intermedia* in the presence of disparlure. *Envirn. Entomol.* 2(3): 485–86.

―――; Schwalbe, C. P.; Beroza, M.; and Knipling, E. F. 1974. Disruption of gypsy moth mating with microencapsulated disparlure. *Sci.* 183: 972–73.

Campbell, R. W. 1963. Some factors that distort the sex ratio of the gypsy moth, *Porthetria dispar* (L.) (Lepidoptera: Lymantrüdae). *Can. Entomol.* 95(5): 465–74.

―――. 1967. The analysis of numerical change in gypsy moth populations. Forest Serv. Mono. 15. 33 pp.

―――. 1970. Problem analysis for population dynamics of the gypsy moth. USDA, Forest Ser. Res. Memo. 46 pp.

―――, and Podgwaite, J. D. 1971. The disease complex of the gypsy moth. I. Major components. *J. Invertebr. Pathol.* 18: 101–7.

Cantwell, G. E.; Dutky, S. R.; Keller, J. C.; and Thompson, C. G. 1961. Results of tests with *Bacillus thuringiensis* Berliner against gypsy moth larvae. *J. Insect. Pathol.* 3: 143–47.

Cardé, R. T.; Roelofs, W. L.; and Doane, C. C. 1973. Natural inhibitor of the gypsy moth sex attractant. *Nature.* 241: 474–5.

Clement, G. E., and Munro, W. 1917. Control of the gypsy moth by forest management. *USDA Bull.* 484.

Cobb, S. S. 1968. Statement of policy for pest control in forestry practices. Speech given at the Senate Committee to Investigate Policies and Laws Relating the Use of Pesticides. July 30.

Cogsburn, R. R.; Tilton, E. W.; and Burkholder, W. E. 1966. Gross effects of gamma radiation of the indian meal moth and angoumois grain moth. *J. Econ. Entomol.* 59: 682–85.

Collins, C. W. 1915. Disperson of the gipsy moth larvae by the wind. *USDA Bull.* 273. 1–23.

————. 1917. Methods used in determining wind dispersion of the gipsy moth and some other insects. *J. Econ. Entomol.* 10: 170–77.

————, and Baker, W. L. 1934. Exploring the upper air for the windborne gipsy moth larvae. *J. Econ. Entomol.* 27: 320–27.

————, and Downey, J. 1967. Laboratory and field evaluation of chemosterilants for gypsy moth in 1964, 1965. *J. Econ. Entomol.* 60(1): 265.

————, and Hood, C. E. 1920. Gipsy moth tree-banding material: how to make, use, and apply it. *USDA Bull.* 899. 18 pp.

————, and Potts, S. F. 1932. Attractants for the flying gypsy moths as an aid in locating new infestations. *USDA Tech. Bull.* 336. 44 pp.

Cosenza, B. J. and Podgwaite, J. D. 1966. A new species of *Proteus* isolated from larvae of the gypsy moth, *Porthetria dispar* (L.). *Antonie van Leeuwenhoek J. Microbiol. Serol.* 32: 186–91.

Craighead, F. C. 1940. Some effects of artificial defoliation on pine and larch. *J. Forest.* 38: 885–88.

Crossman, S. S. 1925. Two imported egg parasites of the gipsy moth, *Anastatus bifasciatus* (disparis) Fonsc. and *Schedius* (Ooencyrtus) *kuwanae* Howard. *J. Agr. Res.* 30: 643–75.

Doane, C. C. 1966. Field tests with newer materials against the gypsy moth. *J. Econ. Entomol.* 59: 618–20.

————. 1968. Aspects of mating behavior of the gypsy moth. *Ann. Entomol. Soc. Am.* 61(3): 768–73.

————. 1969. Trans-ovum transmission of a nuclear-polyhedrosis virus in the gypsy moth and the inducement of virus susceptibility. *J. Invertebr. Pathol.* 14: 199–210.

————, and Cardé, R. T. 1973. Competition of gypsy moth males at a sex-pheromone source and a mechanism for terminating searching behavior. *Environ. Entomol.* 2(4): 603–05.

————, and Schaefer, P. W. 1971. Aerial application of insecticides for control of the gypsy moth with studies of effects on non-target insects and birds. *Conn. Agr. Sta. Bull.* 724. 1–23.

————, and Wallis, R. C. 1964. Enhancement of the action of *Bacillus thuringiensis* var. *thuringiensis* Berliner on *Porthetria dispar* (Linnaeus) in laboratory tests. *J. Insect. Pathol.* 6: 423–29.

Doskotch, R. W. 1973. Chemical factors affecting the feeding behavior of the gypsy moth. Speech given at Gypsy Moth Technical Work Conference. Beltsville, Maryland, February 6–7.

Dowden, P. B. 1961. The gypsy moth eggs parasite, *Ooencyrtus kuwanai*, in southern Connecticut in 1960. *J. Econ. Entomol.* 54:875–78.

Downes, J. A. 1959. The gypsy moth and some possibilities of control of insects by genetical means. *Can. Entomol.* 91:661–64.

Dunbar, D. M., and Doane, C. C. 1973. Gypsy moth and elm spanworm suppression: field evaluation of natural and synthetic pyrethroids. *J. Econ. Entomol.* 66(4):983–86.

———, and Kay, H. K. 1972. *Bacillus thuringiensis:* control of the gypsy moth and elm spanworm with three new commercial formulations. *J. Econ. Entomol.* 65(4):1119–21.

———; Weseloh, R. M.; and Walton, G. S. 1972. A fungus observed on egg clusters of the gypsy moth, *Porthetria dispar* (Lepidoptera: Lymantriidae). *Ann. Entomol. Soc. Am.* 65(6):1419–21.

Eckert, J. E. 1949. The present relation of insecticides to beekeeping. *Am. Bee J.* 89:182.

Farsky, O. 1938. Nonnekontroll und vorbeugungsmethods nach Professor Forst. *Ing. Ant. Anz. Schadlingskunde.* 14:52–65.

Forbush, E. H., and Fernald, 1896. *The gypsy moth.* Boston: Wright and Potter Printing Co. 495 pp.

Forbush, E. H., and Fernald, C. H. 1896. *The gypsy moth.* Technical Work Conference. Beltsville, Maryland. February 6–7.

Fringer, R. 1973. The integrated gypsy moth control program, New Jersey Department of Agriculture. Speech given at Gypsy Moth Technical Work Conference. Beltsville, Maryland. February 6-7.

Gaston, L. K.; Shorey, H. H.; and Saario, C. A. 1967. Insect population control by the use of sex pheromones to inhibit orientation between the sexes. *Nature* (London). 213:1155.

Goldschmidt, R. 1934. *Lymantria. Biblio. Genetia.* 11:1–85.

Golubev, A. V., and Semevsky, F. N. 1969. Distribution of an endemic population of the gypsy moth. (In Russian; English summary). *Zool. Zhur.* 48:850–59.

Granett, J. 1973. A disparlure-baited box trap for capturing large numbers of gypsy moths. *J. Econ. Entomol.* 66(2):359–62.

Grimble, D. G. 1973. Environmental impact of dylox. Speech given at Gypsy Moth Technical Work Conference. Beltsville, Maryland. February 6-7.

Haller, H. L.; Acree, F., Jr.; and Potts, S. F. 1944. Sex attractant of the female gypsy moth. *J. Amer. Chem. Soc.* 66:1659–62.

Hamlen, R. A. 1972. Mass deaths from virus possible control agent for gypsy moth larvae. *Sci. in Agr.* 19(2):3–4.

Hanson, J. B., and Reardon, R. C. 1973. Selected references pertaining to gypsy moth parasites and invertebrate predators. Misc. Bull. USDA, Forest Service, Northeastern Area. Delaware, Ohio. 26 pp.

Hasse, E. 1888. Dufteinrichtungen indischer Schmetterlinge. *Zool. Anz.* 11:475.

Hope, J. G. 1973. Press release no. 73–584. Commonwealth of Pennsylvania, Dep. of Environ. Res. Harrisburg. December 10.

Houston, D. R. 1973. Tree stress and mortality. Speech given at Gypsy Moth Technical Work Conference. Beltsville, Maryland. February 6–7.

Howard, L. O., and Fiske, W. F. 1911. The importation into the United States of the parasites of the gypsy moth and brown-tail moth. *USDA, Bur. Entomol. Bull.* 91. 344 pp.

International Minerals and Chemical Corporation. 1971. Thuricide, microbial insecticide—insect control manual. Libertyville, Illinois. 20 pp.

Jacobson, M. 1960. Synthesis of a highly potent gypsy moth sex attractant. *J. Org. Chem.* 25:29–45.

———. 1965. *Insect sex attractants.* New York: J. Wiley and Sons, Inc. 154 pp.

———. 1966. Masking the effects of insect sex attractants. *Sci.* 154:422.

Jacobson, M., and Jones, W. A. 1962. Insect sex attractants II. The synthesis of a highly potent gypsy moth sex attractant and some related compounds. *J. Org. Chem.* 27:2523–24.

Jankovic, M.; Zecevic, D.; and Vojnovic, V. 1959. Races of the gypsy moth in Yugoslavia. (In Serbo-Croatian; English summary). *Zast. Bilija.* 56:99–107.

Katsanos, J. J. 1973. The gypsy moth survey program—1972. Speech given at Gypsy Moth Technical Work Conference. Beltsville, Maryland. February 6–7.

Klun, J. A. and Robinson, J. F. 1970. Inhibition of European corn borer mating by *cis*-11-tetradecenyl acetate, a borer sex stimulant. *J. Econ, Entomol.* 63:1281–83.

Knipling, E. F. 1968. The role of chemicals in the general insect control picture. *J. Econ. Entomol.* 14:102–6.

———. 1970. Suppression of pest Lepidoptera by releasing partially sterile male: a theoretical appraisal. *Bioscience.* 20(8):465–70.

———, and McGuire, J. U. 1966. Population model to test theoretical effects of sex attractants used for insect control. *Gric. Inform. Bull.* 308. 20 pp.

Koski, J. T. 1971a. Presentation at the Central Plant Board Meeting for USDA, Agr. Res. Serv. Jefferson City, Missouri. February 15–18.

———. 1971b. Guidelines for participation by agricultural research service and forest service in the cooperative gypsy moth action program. Correspondence ER72–24. October 28. 3 pp.

Kozlowski, T. T. 1969. Tree physiology and forest pests. *J. Forest.* 67:118–23.

Kramer, P. J., and Wetmore, T. H. 1943. Effects of defoliation on old resistance and diameter growth of broad-leaved evergreens. *Am. J. Bot.* 30:428–31.

Kulman, H. M. 1965. Effects of artificial defoliation of pine on subsequent shoot and needle growth. *Forest Sci.* 11:90–98.

———. 1971. Effects of insect defoliation on growth and mortality of trees. *Ann. Rev. Entomol.* 16:289–324.

La Fleur, V. A. 1973. Mobile homes, recreational vehicles, and the gypsy moth regulatory procedures. Speech given at Gypsy Moth Techincal Work Conference. Beltsville, Maryland. February 6–7.

Leonard, D. E. 1957. Feeding rhythm in larvae of the gypsy moth. *J. Econ. Entomol.* 65(5):1454–57.

———. 1966. Differences in development of strains of the gypsy moth, *Porthetria dispar* (L.). *Bull. Conn. Agr. Exp. Sta.* 680. 31 pp.

———. 1967. Silking hebavior of the gypsy moth, *Porthetria dispar. Can. Entomol.* 99:1145–49.

————. 1968. Effects of density on larvae on the biology of the gypsy moth, *Porthetria dispar. Ent. Exp. and App.* 11 : 291–304.

————. 1970a. Intrinsic factors causing qualitative changes in populations of *Porthetria dispar.* (Lepidoptera : Lymantriidae). *Can. Ent.* 102 : 239–49.

————. 1970b. Feeding rhythms in larvae of the gypsy moth. *J. Econ. Entomol.* 63 : 1454–57.

————. 1971. Air-borne dispersal of larvae of the gypsy moth and its influence on concepts of control. *J. Econ. Entomol.* 64 : 638–41.

————. 1972. Survival in gypsy moth population exposed to low winter temperatures. *Environ. Entomol.* 1(5) : 549–54.

————. 1973. The potential of Japanese-American intersexes for genetic suppression of gypsy moth populations. Speech given at Gypsy Moth Technical Work Conference. Beltsville, Maryland. February 6–7.

————. 1974. Recent developments in ecology and control of the gypsy moth. *Ann. Rev. Entomol.* 19 : 197–229.

————, and Doane, C. C. 1966. An artificial diet for the gypsy moth, *Porthetria dispar.* (Lepidoptera : Lymantriidae). *Ann. Entomol. Soc. Am.* 59 : 462–64.

Leonard, J. N. 1972. Time-Life Books. In chapter 4 : In man's body, Debts to his past. Life before man. The Emergence of Man Series. 118.

Levesque, G. 1963. A technique for sexing fully developed embryos and early-instar larvae of the gypsy moth. U.S. Forest Ser. Res. Note, NE–2, 3 pp.

Lewis, F. 1973a. Summary of gypsy moth polyhedrosis virus studies—1972. Speech given at Gypsy Moth Technical Work Conference. Beltsville, Maryland. February 6–7.

————. 1973b. Summary of *Bt* tests. Speech given at Gypsy Moth Technical Work Conference. Beltsville, Maryland. February 6–7.

McLane, W. H. 1973. Laboratory screening. Speech given at Gypsy Moth Technical Work Conference. Beltsville, Maryland. February 6–7.

McManus, M. L. 1973. Dispersal model for newly-hatched gypsy moth larvae. Speech given at Gypsy Moth Technical Work Conference. Beltsville, Maryland. February 6–7.

Maksimovic, M. 1958. Experimental research on the influence of temperature upon the development and the population dynamics of the gypsy moth (*Liparis dispar* L.). *Posebna. Izd. Biol. Inst. NR Srb. Beogr.* 3:1–115. (Trans. from Serbo-Croatian 1963. *OTS* 61–11203:1–95).

Marx, H., ed. 1970a. *Gypsy Moth-er Newsletter*, no. 2. 7pp.

———, ed. 1970b. *Gypsy Moth-er Newsletter*, no. 3. 14 pp.

———, ed. 1972a. *Gypsy Moth-er Newsletter*, no. 6. 9 pp.

———, ed. 1972b. *Gypsy Moth-er Newsletter*, no. 7. 9 pp.

———, ed. 1972c. *Gypsy Moth-er Newsletter*, no. 8. 12 pp.

———, ed. 1973a. *Gypsy Moth-er Newsletter*, no. 9. 9 pp.

———, ed. 1973b. *Gypsy Moth-er Newsletter*, no. 11. 10 pp.

———, ed. 1974a. *Gypsy Moth-er Newsletter*, no. 13. 9 pp.

———, ed. 1974b. *Gypsy Moth-er Newsletter*, no. 15. 17 pp.

———, ed. 1975. *Gypsy Moth-er Newsletter*, no. 16. 6 pp.

Merriam, W. A.; Tower, G. C.; Paszer, E. C.; and McDonough, J. L., 1970. Laboratory and field evaluation of insecticides against the gypsy moth. *J. Econ. Entomol.* 63(1):155–59.

Metcalf, C. L.; Flint, W. P.; and Metclaf, R. L. 1962. *Destructive and useful insects, their habits and control*. 4th ed. New York: McGraw-Hill. 1087 pp.

Metterhouse, W. 1973. Gypsy moth parasite and releasing. Speech given at Gypsy Moth Technical Work Conference. Beltsville, Maryland. February 6–7.

———. 1974. Parasite rearing and evaluation. Misc. paper. New Jersey Department of Agriculture. February. 10 pp.

Mikkola, K. 1971. The migratory habit of *Lymantria dispar* (Lepidoptera: Lymantriidae) adult of continental Eurasia in the light of flight to Finland. *Acta. Entomol. Fann.* 28:107–20.

Mosher, F. H. 1915. Food plants of the gypsy moth in America. *USDA Bull.* 250. 39 pp.

Morse, R. A. 1961. The effect of sevin on honey bees. *J. Econ. Entomol.* 54:566–68.

National Academy of Sciences. 1969. Insect-Pest Management, Control and Animal Pest Control, vol. 3, Publication ISBN 0-309-01695-9, Committee on Plant and Animal Pests. Washington, D.C. 508 pp.

Nichols, J. O. 1962. The gypsy moth in Pennsylvania. *Penn. Dep. Agr. Misc. Bull.* 4404. 82 pp.

———. 1971. A major disaster for Pennsylvania? *Penn. Forest Pest Rep.* 47. Penn. Dep. Environ. Res. Harrisburg. 3 pp.

———. 1972. *Penn. Forest Pest Rep.* 53. Penn. Dep. Environ. Res. Harrisburg. 4 pp.

———. 1973. Guidelines for county-state-federal cooperative gypsy moth control in high-use areas. Speech given at Gypsy Moth Technical Work Conference. Beltsville, Maryland. February 6–7.

———. 1976. Pest Report 76–3. Penn. Dep. Environ. Res. Harrisburg. 3 pp.

O'Brien, R. D., and Wolfe, L. S. 1964. *Radiation, radioactivity, and insects.* New York: Academic Press. 193 pp.

O'Dell, T. M. 1973. Gypsy Moth Report—Untitled. Speech given at Gypsy Moth Technical Work Conference. Beltsville, Maryland. February 6–7.

———, and Rollinson, W. D. 1966. A technique for rearing the gypsy moth. *Porthetria dispar* (L.), on an artificial diet. *J. Econ. Entomol.* 59 : 741–42.

O'Neil, L. C. 1963. The suppression of growth rings in jack pine in relation to defoliation by the Swaine jack-pine sawfly. *Can. J. Bot.* 41 : 227–35.

Pantyukhov, G. A. 1964. The effect of negative temperatures upon different populations of *Euproctis chrysorrhoea* L. and *Lymantria dispar* L. (In Russian; English summary). *Entomol. Obozr.* 43 : 94–111. (*Rev. Appl. Entomol.* A 54 : 434).

Patil, G. P., Stiteler, W. M. 1973. A brief report on the gypsy moth sampling study. Speech given at Gypsy Moth Technical Work Conference. Beltsville, Maryland. February 6–7.

Pennsylvania Department of Environmental Resources. 1973. *Reasons Why Massive Chemical Spraying Programs to Combat the Gypsy Moth Are Not Recommended.* Bur. Forestry, Div. Forest Pest Man. Harrisburg. 2 pp.

———. 1975. Forestry, Programs and Services. Harrisburg. 45 pp.

———. 1976. 1975 Annual Report of Forest Insect and Disease Conditions in Pennsylvania. Harrisburg. 34 pp.

Perry, C. C. 1955. Gypsy moth appraisal program and proposed plan to prevent spread of the moths. *USDA Tech. Bull.* 1124. 27 pp.

Podgwaite, J. D. 1973. Disease diagnosis and evaluation in natural gypsy moth populations: progress to date. Speech given at Gypsy Moth Technical Work Conference. Beltsville, Maryland. February 6–7.

———, and Campbell, R. W. 1972. The disease complex of the

gypsy moth. II. Aerobic bacterial pathogens. *J. Inverte. Pathol.* 20(3): 303–8.

Quimby, J. W. 1972. Tree mortality damage, appraisal in Pike and Monroe Counties—1972. Misc. paper. Penn. Dep. Environ. Res. Harrisburg. 1 pp.

————. 1973. Gypsy moth parasite releases and evaluation. Pennsylvania Department of Environmental Resources 1973 annual report of operations and forest pest conditions. Harrisburg. 7–11.

Richerson, J. V. 1973. Sexual behavior of the gypsy moth. Speech given at Gypsy Moth Technical Work Conference. Beltsville, Maryland. February 6–7.

Rollinson, W. D.; Lewis, F. B.; and Waters, W. E. 1965. The successful use of a nuclear-polyhedrosis virus against the gypsy moth. *J. Inverte. Pathol.* 7: 515–17.

Rose, A. H. 1969. Noteworthy forest insects in Ontario in 1969. *Proc. Entomol. Soc. Ont.* 100: 11–13.

Rule. 1961. Gamma irradiation effects on spermatogenesis in the gypsy moth. Quarterly Report, October — December. Northeastern Forest Exp. Sta. Upper Darby, Pennsylvania.

Sailer, R. I. 1973. Foreign work on gypsy moth parasites during 1972. Speech given at Gypsy Moth Technical Work Conference. Beltsville, Maryland. February 6–7.

Sanders, H. J. 1975. New weapons against insects. *Chem. and Eng. News* 53(30): 18–31.

Sandquist, R. E.; Richerson, J. V.; and Cameron, E. A. 1973. Flight of North American female gypsy moths. *Environ. Entomol.* 2(5): 957–58.

Shorey, H. H.; Gaston, L. K.; and Saario, C. A. 1967. Sex pheromone of noctuid moths. 14. Feasibility of behavioral control by disrupting pheromone communication in cabbage loopers. *J. Econ. Entomol.* 60: 1541–45.

————; Kaae, R. S.; Gaston, L. K.; and McLaughlin, J. R. 1972. Sex pheromone of Lepidoptera. 30. Disruption of sex pheromone communication in *Trichoplusia ni* as a possible means of mating control. *Environ. Entomol.* 1: 641–45.

Simons, E. E. 1973. Status of other forest insects. Pennsylvania Department of Environmental Resources 1973 annual report of operations and forest pest conditions. Harrisburg. 12–15.

Smilowitz, Z., and Rhoads, L. D. 1972. Parasites of gypsy moth located in Pennsylvania. *Sci. Agr.* 19(2):5.

————. 1973. An assessment of gypsy moth natural enemies in Pennsylvania. *Environ. Entomol.* 2(5):797–99.

Smulyan, M. T. 1923. The barrier factors in gipsy moth tree-banding material. *USDA Bull.* 1142. 16 pp.

Snodgrass, R. E. 1935. *Principles of insect morphology.* New York: McGraw-Hill. 667 pp.

Solomon, M. E. 1949. The natural control of animal populations. *J. Anim. Ecol.* 18:1–34.

Stefanov, D. and Kermidchiev, M. 1961. The possibility of using some predators and parasitic insects (entomophagous insects) in the biological control of the gypsy moth (*Lymantria dispar* L.). (Bulgarian). *Nauch. Tr. Vissh. Lesotekh. Inst.* 9:159–68. (*Rev. Appl. Entomol.* A 53:170).

Stephens, G. R. 1971. The relation of insect defoliation to mortality in Connecticut forests. *Conn. Agr. Exp. Sta. Bull.* 723. 16 pp.

————; Turner, N. C.; and De Roo, H. C. 1972. Some effects of defoliation by gypsy moth (*Porthetria dispar* L.) and elm spanworm (*Ennomos subsignarius* Hbn.) on water balance and growth of deciduous forest trees. *Forest Sci.* 18(4):326–30.

Stevens, L. J. 1973a. Section IV, disparlure field tests. Speech given at Gypsy Moth Technical Work Conference. Beltsville, Maryland. February 6–7.

————. 1973b. Section VII, sterile male technique, rearing and field tests. Speech given at Gypsy Moth Technical Work Conference. Beltsville, Maryland. February 6–7.

Strang, G. E.; Nowakowski, J.; and Morse, R. A. 1968. Further observations on the effect of carbaryl on honey bees. *J. Econ. Entomol.* 61:1103–4.

Sullivan, C. R., and Wallace, D. R. 1972. The potential northern dispersal of the gypsy moth, *Porthetria dispar* (Lepidoptera: Lymantriidae). *Can. Entomol.* 104:1349–55.

Summers, J. N. 1922. Effects of low temperature on the hatching of gypsy moth eggs. USDA Bull. no. 1080. 54 pp.

Tigner, T. 1973. Cooperative survey of gypsy moth parasites in New York State. Speech given at Gypsy Moth Technical Work Conference. Beltsville, Maryland. February 6–7.

Tomlin, A. D., and Forgash, A. J. 1972a. Toxicity of insecticides to gypsy moth larvae. *J. Econ. Entomol.* 65:953–54

———. 1972b. Penetration of gardona and DDT in gypsy moth larvae and house flies. *J. Econ. Entomol.* 65(4) : 942–45.

United States Department of Agriculture. 1957. *U.S.D.A. Launches Large-scale Effort to Wipe Out Gypsy Moth.* News Release. March 30. 7 pp.

———. 1973c. Map : *Gypsy Moth Survey 1973.* USDA, APHISfi PP&Q.

———. 1973a. Map : *Gypsy Moth Spread, 1869–1973.* USDA, APHIS, PP&Q.

———. 1973b. Map : *Gypsy Moth Defoliated Areas, 1973.* USDA, APHIS, PP&Q.

———. 1973c. Map : *Gypsy Moth Suveys 1973.* USDA, APHIS, PP&Q.

———. 1973d. *Gypsy Moth, a Major Pest of Trees. Bull.* PA–1006. 12 pp.

Uvarov, B. P. 1921. A revision of the genus *Locusta* L. (-*Pachylylus Fieb.*) with a new theory as to the periodicity and migration of locusts. *Bull. Ent. Res.* 12 : 135–63.

Valentine, H. T. 1973. The role of forest simulators in gypsy moth research. Speech given at Gypsy Moth Technical Work Conference. Beltsville, Maryland. February 6–7.

Vailijevic, L. 1958. Share of polyhedra and other diseases in reduction of the gypsy moth that took place in the P. R. of Serbia in 1957. (In Serbo-Croatian; English Summary). *Zast. Bilja.* 41–2 : 123–37.

Waggoner, P. E., and Turner, N. C. 1971. Transpiration and its control by stomata in a pine forest. *Conn. Agr. Exp. Sta. Bull.* 726. 87 pp.

Wallis, R. C. 1957. Incidence of polyhedrosis of gypsy moth larvae and the influence of relative humidity. *J. Econ. Entomol.* 50 : 580–83.

———. 1962. Environmental factors and epidemics of polyhedrosis in gypsy moth larvae. *Proc. Int. Congr. Entomol.,* 11th. Wien. 2 : 150–53.

Warner, E. G. 1973. Regulatory and control. Speech given at Gypsy Moth Technical Work Conference. Beltsville, Maryland. February 6–7.

Waters, R. M., and Jacobson, M. 1965. Attractiveness of gyplure masked by impurities. *J. Econ. Entomol.* 63 : 943–45.

Waters, R. M., and Rule. 1960. Further investigations of gamma irradiation of the gypsy moth. Quarterly Report, October–

December. Northeastern Forest Exp. Sta., Div. Forest Insect Res. Upper Darby, Pennsylvania. 1–6.

Weseloh, R. M. 1972. Influence of gypsy moth egg mass dimensions and microhabitat distribution on parasitization by *Ooencyrtus kuwanai*. *Ann. Entomol. Soc. Am.* 65(1):64–9.

———. 1973. Relationships of natural enemy field populations to gypsy moth abundance. *Ann. Entomol. Soc. Am.* 66(4): 853–56.

White, W. B. 1973. Socioeconomic impacts. Speech given at Gypsy Moth Technical Work Conference. Beltsville, Maryland. February 6–7.

Wright, R. H. 1963. Insect control by nontoxic means. *Sci.* 144:487.

Yendol, W. G.; Hamlen, R. A.; and Lewis F. B. 1973. Evaluation of *Bacillus thuringiensis* for gypsy moth suppression. *J. Econ. Entomol.* 66(1):183–86.

Zanforlin, M. 1970. The inhibition of light orientating reactions in caterpillars of Lymantriidae, *Lymantria dispar* (L.) and *Orgyi antiqua* (L.). *Monit. Zool. Ital.* 4:1–19.

Index

231